Supernovae

SUPERNOVAE

Paul Murdin
Royal Greenwich Observatory

Lesley Murdin
The Open University

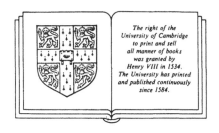

The right of the
University of Cambridge
to print and sell
all manner of books
was granted by
Henry VIII in 1534.
The University has printed
and published continuously
since 1584.

CAMBRIDGE UNIVERSITY PRESS

Cambridge
London   New York   New Rochelle
Melbourne   Sydney

To our parents

*That tho' a man were admitted into heaven
to view the wonderful fabrick of the world,
and the beauty of the stars, yet what would
otherwise be rapture and extasie, would be
but a melancholy amazement if he had not a
friend to communicate it to.*

Archytas (attr.)

CAMBRIDGE UNIVERSITY PRESS
Cambridge, New York, Melbourne, Madrid, Cape Town,
Singapore, São Paulo, Delhi, Tokyo, Mexico City

Cambridge University Press
The Edinburgh Building, Cambridge CB2 8RU, UK

Published in the United States of America by Cambridge University Press, New York

www.cambridge.org
Information on this title: www.cambridge.org/9780521189798

First published in 1978 by Reference International Publishers Ltd as
The new astronomy and © Reference International Publishers 1978

This revised edition published in 1985 by Cambridge University Press as
Supernovae and © Cambridge University Press 1985

First paperback edition 2011

A catalogue record for this publication is available from the British Library

Library of Congress Cataloguing in Publication data

Murdin, Paul.
    Supernovae.
    Rev. ed. of: The new astronomy. C1978.
    Bibliography: p.
    Includes index.
    1. Supernovae.   1. Murdin, Lesley.   11. Murdin, Paul.
New astronomy.   111. Title.
QB43.S95M87 1985   523   84–23833

ISBN 978-0-521-30038-4 Hardback
ISBN 978-0-521-18979-8 Paperback

Contents

Preface

We wrote this volume because we wanted to tell an optimistic story. We wanted to tell how, from isolated events noted in old manuscripts, smudged on photographic plates, and clocked photon by photon, mankind has pieced together evidence about the place of supernovae in the scheme of things.

Poets, geologists, biologists, physicists and, of course, astronomers have had their different views of the significance of supernovae. They have used them in their own way: as a spark for their imagination, as a sidelight to their main study, as a laboratory for conditions which can never be achieved on Earth, and as a subject on which to spend a lifetime's study. They did so with all the range of human reactions, but what we perceive as the common thread in the story is their enjoyment of it all. It shows in small ways and large: the sly names for new things, like the *Urca process* and *Geminga*; the exhilaration from an unexpected discovery like Bell's discovery of pulsars; the regret when an astronomer realizes he has gone down the wrong path, as in Rosse's note in his journal for 1848 November 29; the satisfaction which a group feels when together they have worked on and solved a difficult puzzle, like the tracking down of the millisecond pulsar. Enjoyment shows in the present life of astronomers: their broad grins at midday breakfast when the sky has been clear the night before; their banter on the bus at a conference outing. Even their crestfallen look at the end of an observing run when the equipment hasn't worked, the tension in a conference room when they stop just this side of saying 'I don't believe you' and the bickering about recognition and prizes – they all show how strongly astronomers feel about their work. If despair is deep, then elation will be high when it comes. The confidence is there, with the self-knowledge that it is difficult to understand everything and that it is marvellous to understand anything at all. The study of supernovae represents, for us and in this volume, a part of astronomy, a part of science, a part of knowledge and a part of human life.

Paul and Lesley Murdin
September, 1984

1

Supernovae in space and time

From time to time bright new stars called supernovae flare briefly in the sky. In fact, there have been only five supernovae seen by the unaided eye in the last 1000 years. But supernovae have an importance in astronomy which transcends their mere numbers. Why?

The Universe is sending us at this moment virtually all the information that it ever will send us. Anything that we can find out about the Universe we can probably find out now. Understanding does not come easily. The Universe does not arrange itself to make itself comprehensible. All the pieces of all the cosmic jigsaws are there, but scrambled and mixed up with scraps which have no significant place in that picture. Supernovae are a large and significant piece of the jigsaw.

But even fundamental clues to the structure of the Universe can be overlooked or unrecognized when seen by an uncomprehending eye. For example the temporary appearance for many months in AD 1006 of a supernova which shone so brightly that it cast shadows on the ground apparently did not change the beliefs of those who thought that the heavens never changed. However, two similar bright supernovae in 1572 and 1604 threw light into the corners of minds ready to understand that the stars were not permanent, and the new astronomy began.

At first astronomers concentrated on determining only the motions of the stars and gave fleeting attention to determining their life cycle. This was because although we can talk of stars being born, living and dying, they go through these stages on a very long time scale and changes are not usually apparent. The Sun has been as it is for some four billion* years. Over a human lifetime, indeed throughout human

existence, most stars *have* remained very much the same. But recognizing from the occasional appearance of so-called new stars that the heavens do change, astronomers have sought to understand how stars evolve. It has not been easy.

Astronomers see very many different kinds of stars in the sky; their objective in looking at these stars is, first, to explain how each kind works and what gives it the appearance that it has to us and, second, to see how it changes from one stage to another.

Astronomers have been compared with Martians presented with a snapshot of a forest. The Martians must try to understand from their photograph the life cycle of a tree. They need to know how an acorn turns into an oak sapling, and how an oak sapling turns into a fully mature tree, how the tree dies and decays on the forest floor. Without actually seeing the processes of change in a tree, the Martians must study the different organisms that they find and try to classify them into their different categories, guessing how they relate to one another.

In the same way, astronomers study the various types of stars, try to classify them under different headings, and try to explain how one type of star changes to another. Remarkably, after only 100 years of studying the intrinsic nature of stars, astronomers now believe that they do have a workable system for explaining how one kind of star changes into another. They believe that, within a general outline, they have explained the whole life cycle of typical stars from how they are born and begin to shine, up to the moment when all their energy is used up and they cease shining brightly and then die.

It is only very recently that astronomers have begun to understand the death of stars. Unlike the major part of a star's life, which is long and peaceful, the stage which marks the death of a

---

* Throughout this volume 1 billion equals 1 thousand million.

star can be brief and dramatic – so brief that it may last just a few months and so dramatic that for a few days the star outshines all of its millions of neighbours put together. The lifetime of stars is so long and this death so brief that, on average, just one such event in a galaxy of 100 billion stars can be witnessed during a working human lifetime.

The bright explosion which marks the death of a star is a supernova. Where no star was seen before, astronomers see a bright new star shining. When first recognized, such objects were named *novae* (pronounced 'no-vee') meaning *new stars*, although astronomers now understand that the star is not new but was previously so faint that it was not noticed before. Just as witnessing a tree fall would add to the Martians' understanding of forests, so witnessing supernovae adds to astronomers' comprehension of the Universe.

In 1935 it was formally understood that there were at least two very distinct kinds of nova, one much brighter than the other. The fainter kind is probably caused by a relatively weak explosion on one of a pair of stars orbiting each other. The brighter kind (which is 100 000 times brighter) is a supernova; it is this explosion which marks a stellar death.

In our own Galaxy of 100 billion stars the records survive of five supernovae in the last 1000 years (none since the invention of the telescope in 1608). But astronomers have found in the sky the remains not only of these supernovae but of others which occurred many thousands of years ago. A supernova explosion produces two visible kinds of objects. At the site where a supernova has occurred, astronomers can see the shell of the exploding star speeding into space in fragments, colliding with tenuous gas in space and glowing from the force of the collision. This shell is called a supernova remnant. At the centre of a

supernova remnant may sit the hard core of the star that has died, a star so faint that most emit no discerned light, a star packed so tightly by the force of the explosion that a matchbox full would weigh a billion tons. It is called a neutron star.

The most-studied example of a supernova and its remnant is known as the Crab. Seen as a bright star in AD 1054, the Crab supernova produced a nebula which was discovered in the eighteenth century. At the centre of the Crab Nebula lies a faint star, the neutron star produced by the supernova. The star is spinning on its axis at a rate of 30 revolutions per second (more than a million times faster than the Earth which rotates once per day). A 'hot spot' on the neutron star shines like a lighthouse into space and, because the beam passes across the Earth once during each revolution of the neutron star, it is perceived to flash or pulse. It is a pulsar.

The study of supernovae has shed light on other unexplained problems in astronomy, such as how the elements came to be formed (including those in our bodies), and on cosmic rays which are speeding subatomic particles in space.

Astronomers believe that supernovae are at the origin of cosmic rays and thus at the origin of part of the natural level of radioactivity on Earth. Some even speculate that past supernovae have played a part, through increasing radioactivity due to the cosmic rays, in the evolution of life itself.

Thus supernovae, worth studying in their own right, have wide-ranging links with other studies as well as occupying a central position in the science of astronomy. Astronomers find it worthwhile to spend time studying not only supernovae in other galaxies (and the remnants of supernovae which have occurred in this Galaxy), but to delve into the tantalizing historical records of past galactic supernovae, attempting to discover the galactic supernovae which caused the remnants that they see.

Research into supernovae is only in small part a matter of library study. Most information comes from investigating the sky. Practically everything has been achieved that can be achieved just by looking, and since the Universe is sending us all the information it ever will, it is only by new techniques that astronomers can achieve any new understanding. Some of the advances come from building bigger and better telescopes to perceive the light from fainter stars with finer detail. But the optical astronomer looks at the Universe through a restricted window in the atmosphere of the Earth. Cosmic ultraviolet light cannot be seen, since it is absorbed by ozone in the atmosphere; cosmic infrared light cannot easily be seen since this is absorbed by water vapour and oxygen. Just before World War II, a second window on the Universe was opened: the window penetrated by radio telescopes. The brightest 'star' seen by radio astronomers turned out to be a supernova remnant formed just 300 years ago by an unseen supernova. Among the other bright radio 'stars' is Taurus A, the radio astronomer's name for the Crab Nebula, remnant of the supernova of 1054.

Among the fainter radio stars were discovered the pulsars, now known to be neutron stars formed in supernova explosions.

No further wide windows onto the Universe are available to the Earth-bound observer. To see the Universe of stars through other windows, the astronomer flies telescopes above the atmosphere. The first X-ray 'star' to be identified, seen by a rocket-borne X-ray telescope, was the Crab Nebula. But it has been the previously unknown X-ray stars which have been the greatest surprise: the so-called compact X-ray stars are neutron stars (such as the one in the Crab) and – more bizarre – black holes, also formed in supernova explosions.

The contemporary study of astronomy has been described by Geoffrey Burbidge as being divided into the study of the Crab Nebula and the study of everything else. An exaggeration of course, but this remark spotlights the central place held by the Crab and other supernovae in astronomy. This book begins with the search for the historical supernovae and, especially, for the progenitor of the Crab Nebula, the supernova of 1054.

Guest stars

In the first year of the period Chih-ho, the fifth moon, the day chi-ch'ou, a guest star appeared approximately several inches southeast of Thien-kuan. After more than a year it gradually became invisible.

In this plain statement Toktaga and Ouyang Hsuan, the fourteenth century Chinese authors of the official history of the Sung dynasty (the *Sung Shih*), noted the appearance of a previously unknown bright star in the constellation now known as Taurus, the Bull. The day referred to is what we would now call 1054 July 4; the star Thien-kuan is what present-day astronomers called Zeta Tauri, possibly including the few stars around it.* These prosaic details pin down the precise day on which occurred an astronomical event whose effects are still with us over 900 years later.

The guest star of 1054

To the Chinese, guest stars were well worth noting and, indeed, looking out for. They believed that humans lived on Earth in a kingdom roofed with stars, and that human destiny was subject to a 'cosmic wind'. Chinese emperors appointed court astrologers who watched the sky to ascertain the direction in which this cosmic wind would blow their subjects. These astrologers had been noting down celestial events since the fourteenth century BC; these events were believed to mark events of great significance in earthly affairs (such as the death of princes).

FIG. 1. *Chinese imperial astronomers. The Hsi and Ho brothers receive their commission from the Emperor Yao to organize the calendar and pay respect to the celestial bodies. Although Hsi and Ho were legendary, this late Ching representation of the court presentation illustrates the status of astronomers in China. Franz Kühnert wrote in 1888: 'Probably another reason why many Europeans consider the Chinese such barbarians is on account of the support they give to their astronomers – people regarded by our cultivated Western mortals as completely useless. Yet there they rank with Heads of Departments and Secretaries of State. What frightful barbarism!' This figure reproduced from J. Needham,* Science and Civilisation in China, *Cambridge University Press.*

---

* The present system of naming bright stars uses the Latin constellation name and a Greek letter, with Alpha (α) usually denoting the brightest, Beta (β) the second brightest and so on. Zeta Tauri is therefore the sixth brightest star in the constellation of Taurus, the Bull.

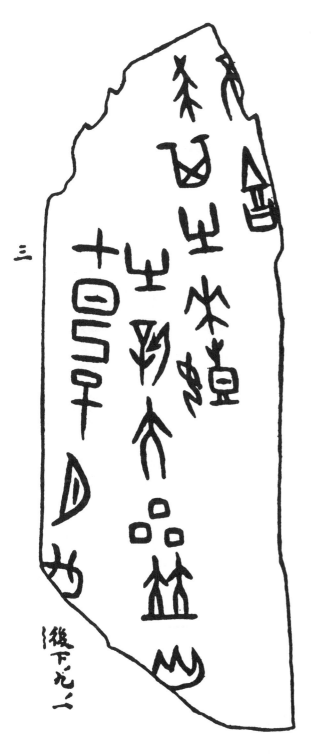

FIG. 2. (*Left*) *Oldest record of a guest star. Chinese oracle bones were made from an animal's shoulder blade and inscribed with a question. After searing the bone with a red hot poker, the answer to the question was divined from the pattern of cracks which resulted. The appearance of a guest star or other cosmic event may have been held to confirm the answer. This bone dates from* 1300 BC *and reads: 'On the 7th day of the month a great new star appeared in company with Antares.' This is the first record of a guest star. Since it was remarked as 'great' it may have been a supernova. This figure reproduced from J. Needham,* Science and Civilisation in China, *Cambridge University Press.*

FIG. 3. (*Above*) *Korean Observatory. In the manner of Chinese astronomers, Korean astronomers occupied the platform at the top of a tower observatory. This one is 30 feet high and was built at Kyungju in the seventh century* AD. *The observatory also had a large window, facing north, for observations of the circumpolar stars. Presumably the function of the tower was to give an uninterrupted view over nearby buildings and trees. This figure reproduced from J. Needham,* Science and Civilisation in China, *Cambridge University Press.*

The observations were noted down not as incidental asides but as part of a deliberate policy. According to the Jesuit Lecompte's account in AD 1696 of the Ch'ing astronomical bureau:

They still continue their observations. Five mathematicians spend every night on the tower watching what passes overhead. One gazes towards the zenith, another to the east, a third to the west, the fourth turns his eyes southwards and a fifth northwards, that nothing of what happens in the four corners of the world may escape their diligent observation. They take notice of the winds, the rain, the air, of unusual phenomena such as eclipses, the conjunction or opposition of planets, fires, meteors and all that may be useful. This they keep a strict account of, which they bring in every morning to the Surveyor of Mathematics, to be registered in his office.

Many of these accounts, religiously kept, have survived in official Chinese histories where they are used as background material, illuminating and justifying imperial actions. For modern historians they illustrate the social history of China; for astronomers they give details about supernovae whose remnants may still be seen in our Galaxy.

The medieval Chinese historian Chang Te-hsiang writes in the *Sung hui-yau* that, soon after the guest star of 1054 became visible, the Director of the Astronomical Bureau, Yang Wei-te, presented himself prostrate and kow-towing before his Emperor to tell him of its appearance. Perhaps Yang was fearful of not having foretold the coming of the guest star, since he assured the Emperor that because the star did not conflict with the constellation Pi (the nearby Hyades star cluster) and was bright and lustrous, it meant that a person of great wisdom and virtue was to be found in that part of China. This oblique compliment was no doubt well received by the

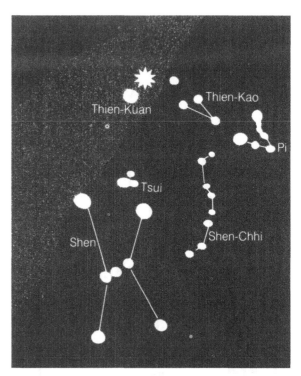

FIG. 4. *Chinese constellations near Taurus. The supernova of 1054 occurred in the Milky Way near the constellation Thien-Kuan, north of Shen, the present constellation of Orion. The supernova 'did not conflict with Pi [the Hyades]', noted one astrologer, reassuringly. In this figure, the supernova is drawn to the northwest of Thien-Kuan, although one record, the Sung Shih, says that it appeared to the southeast. For the reason see page 66.*

Emperor and his assembled court, which acclaimed Yang's remarkably accurate prognostication. Yang requested that it be filed at the Bureau of Historiography, perhaps the better to be retrieved if he should displease the Emperor at a later date and need to re-convince him of his loyalty. The account of this episode goes on:

During the third month in the first year of the Chia-yu reign period [1056 March 19 to April 17] the Director of the Astronomical Bureau reported 'The guest star has become invisible, which is an omen of a guest's departure.' Originally, during the fifth month in the first year of the Chih-ho reign period, the guest star appeared in the morning in the east, guarding Thien-kuan. It was visible in the day, like Venus, with pointed rays in every direction. The colour was reddish white. It was seen like that for twenty-three days altogether.

The guest star was also seen and recorded by Japanese astronomers. They noted its appearance 'in the orbit of Orion', as bright as the planet Jupiter, in early June 1054. Therefore they may have observed it before it was at its brightest (as bright as the planet Venus) which occurred early in July 1054 when it was discovered by Yang Wei-te. Possibly this is why he seemed somewhat apprehensive when making his report to the Emperor. He may have feared being criticized for not noticing it early enough and he may well have had a real danger to fear – according to legend, the Chinese astronomers Ho and Hsi were beheaded for failing to predict the solar eclipse of 2137 BC.

European sightings – or their absence

There are no European records of the guest star. One of the puzzles about the ancient observations of the supernova of 1054 is the question as to why a new star which was as bright as Venus, visible in daylight for 3 weeks, and visible in the night sky for 2 years, should go unrecorded in European annals. Supernovae were neither totally unprecedented nor unknown in historical sources available in Europe. There are two examples.

In the work of two Roman chroniclers there

appear to be references to a supernova recorded by the Chinese in AD 185. In a description of portents and evil omens, Herodias (*c.* AD 250) referred to stars that 'shone continuously by day', and the reference in his history seems to be to the year AD 185. A fourth century history of the reign of the Emperor Commodus, the *Historia Augusta*, says that the heavens were 'ablaze' just before a civil war, again in what appears to be a reference to AD 185. These asides are vague and give little indication of how bright the supernova was, and none at all of where it appeared. Their meaning to later readers in the Dark Ages of Europe was obviously obscure.

Another supernova which was seen in Europe took place in AD 1006. The *Pars Altera* (919–1044) of the chronicle of the Benedictine Abbey at St Gall in Switzerland describes the star, seen low on the southern horizon. It would not have been visible from much further north; if word of it reached northern Europe, it was discounted, disbelieved or at least forgotten.

The supernova of 1054, however, was much more likely to have been seen. It appeared in a recognizable part of the sky and it passed near the zenith as seen from Europe, yet it was not noted. Some writers blame the cloudy weather. British astronomers in particular are ready to accept this explanation as a reason for the star not having been recorded. But was the weather over the whole continent, the Mediterranean and North Africa persistently bad for many months? In all probability the supernova *was* seen; we are still left with the question as to why it was not recorded or why the records have not survived.

One possible explanation that has been advanced for the lack of records of this supernova is that its appearance coincided with the split in Christianity between the Roman Catholic Church to the west and Greek Orthodox Church in the

٢٤٢

'One of the well known epidemics of our own time is that which occurred when the spectacular star appeared in Gemini in the year 446 H. In the autumn of that year fourteen thousand people were buried in the cemetery of the church of St Luke, after all the other cemeteries in Constantinople had been filled. Then, in midsummer the Nile was low, and most people in Old Cairo and all the strangers died, except those whom Allah willed to live. The epidemic spread to Iraq and affected most of the population, and the land was laid waste in the wake of contending troops, and this continued until the year 454 H. In most countries people fell ill with black-bile ulcers and swelling of the spleen. The usual arrangement of the rise and fall of the fevers was altered and the order of the crises was upset, so that the rules of prognosis had to be changed. As this spectacular star appeared in the sign of Gemini, which is the ascendent of Egypt, it caused the epidemic to break out in Old Cairo when the Nile was low, at the time of its appearance. Thus Ptolemy's prediction became true: Woe to the people of Egypt when one of the comets appears threateningly in Gemini! Then when Saturn descended into the sign of Cancer, the destruction of Iraq, Mosul and Jazira was completed; Diyar Bakr, Rabica, Mudar, Fars and Kirman, the lands of the Maghrib, Yemen, Fustat and Damascus/Syria were upset; the affairs of the kings of the world were disturbed; and wars, famine and epidemics abounded. And this confirmed the wisdom of Ptolemy in saying: when Saturn and Mars are in conjunction in the sign of Cancer, the world will be shaken.'

FIG. 5. *Arabic reference to the Crab supernova. Page 242 of this edition of Ibn Abi Usaybia's medical textbook* Uyun al-Anba fi tabaqat al-Atibba [*Important information concerning the generations of physicians*] *of AD 1242 refers to the supernova of 1054. It quotes Ibn Butlan, a physician of Baghdad. The whole quotation illustrates the general upset to order which Arabic astrologers connected with the appearance of comets and novae, and with the occurrence of conjunctions of planets in zodiacal constellations. Astrology was at this time an important adjunct to the practice of medicine.*

east. Events were leading up to the split between July 16 and 24, 1054, when the formal break occurred between the patriarch Michael Cerularins and Pope Leo IX. This was the beginning of the rift which became known as the Great Schism. The Church Fathers might have found it expedient to expunge from history such a dramatic portent. At a time of political instability, it would have been a brave person who risked placing on record an interpretation of this event.

In the Middle East, the single known probable report of the 1054 supernova is connected with an interest in astrology. A medical textbook, *Uyun al-Anba*, composed by Ibn Abi Usaybia in about AD 1242 contains a statement by Ibn Butlan (a Christian physician from Baghdad, who had lived in Cairo and in Constantinople between 1052 and 1055). Butlan said:

One of the well-known epidemics of our time is that which occurred when the spectacular star appeared in Gemini in the year 446 AH.

The year in question lasted from 1054 April 12 until 1055 April 1. The startling reference to Gemini rather than to Taurus can be understood as a reference to the astrological sign which, because of the precession of the equinoxes, was at that time in the constellation Taurus.* Thus the date and position, and the general circumstances of the observation suggest that Butlan was ascribing to the supernova of AD 1054 the cause of a noteworthy plague (which, by his account, spread from Constantinople where 14 000 people perished, to Egypt where it killed most of the population of Cairo, and to Iraq, the Yemen and Syria). Butlan's explanation is in general agreement with the belief that diseases were influences from the stars; the word *influenza* (from the Italian) still preserves this belief in European languages as an etymological fossil.

Save for Butlan's aside, there are no other definite western records of the AD 1054 supernova in Europe. The records are silent about an event which must have been an astronomical spectacle, far outshining such pale rivals as Halley's Comet, which appeared 12 years afterwards and was commemorated in the Bayeux Tapestry.

---

* As the Earth spins daily on its axis, it slowly wobbles like a top (it *precesses*), so that its equator moves relative to the celestial sphere and, therefore, to the ecliptic. The period of precession is 26 000 years. The signs of the zodiac lie along the ecliptic and their positions are measured relative to the place where the Earth's equator cuts the ecliptic. Because the equator moves, the signs of the zodiac move. Originally they were correlated with the constellations, but the signs of the zodiac have drifted so that none of them bears any relation to the constellations after which they were originally named. Thus the zodiacal sign of Aries has passed completely across the constellation Pisces and is now entering the constellation Aquarius. In 1054, the part of the ecliptic labelled with the zodiacal sign of Gemini was in the constellation Taurus; it would be natural for an astrologer, noting an event in the constellation, to describe it as in the zodiacal sign.

The star in an Arizona cave

In North America, pictographs have been found which suggest that the new star actually was seen and recorded there. The evidence was uncovered in the early 1950s by two astronomers on a non-astronomical expedition to rescue cultural relics in northern Arizona before the completion of a dam flooded a valley of the Colorado River.

The astronomers were Helmut Abt, then at Yerkes Observatory, Wisconsin, and Bill Miller, who was chief photographer at Mt Palomar Observatory, California. On their expedition they found two remarkable prehistoric drawings.

The first representation, a drawing in red

FIG. 6. *Supernova of 1054 at White Mesa. The crescent Moon is painted in red haemetite on the sandstone wall of a cave at White Mesa, Arizona. Off the lower cusp, in this photograph by William C. Miller made in 1952, is a circle representing the supernova of 1054.*

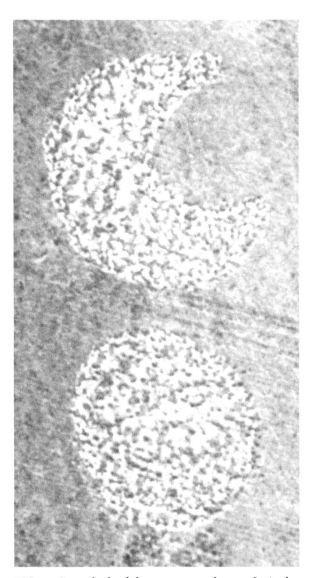

FIG. 7. *Petroglyph of the supernova of 1054. Incised on sandstone at Navajo Canyon, Arizona, a representation of the supernova of 1054 has been rendered visible with chalk in this photograph by Robert Euler. Photograph supplied by William C. Miller.*

pigment on the wall of a cave at White Mesa, shows a crescent Moon with a circle, apparently representing a bright star, overlapping the lower cusp or horn. The second also shows the Moon as a crescent (but reversed from the first) and the circular object is directly under the lower cusp. This representation is a petroglyph, incised on a canyon wall near ruins in Navajo Canyon.

Because the first drawings were found in association with Pueblo Indian dwellings it is possible to give some indication of their date by archaeological methods. Pueblo Indian pottery from different locations and of different dates is highly distinctive. Potsherds collected at the two sites have been analysed by Robert C. Euler, Curator of Anthropology at the Museum of Northern Arizona; he found that most of the inhabitants of the White Mesa site lived there later than AD 1070, but that there were earlier sherds in the collection. At the Navajo Canyon site a deep arroyo (gulley), cut into the floor of the canyon, had exposed broken pieces of pot, with the oldest at the deeper levels. Euler determined that the site was occupied by Pueblo Indians from before AD 700 to after AD 1300 with about one fifth of the sherds collected dating from between AD 900 and 1100.

Astronomical themes are rarer in American Indian art than, say, hunting themes, but they are not unknown. Well-established depictions of astronomical events occur on historically authenticated calendars or so-called winter counts by the Dakota Sioux. The Leonid meteor shower marks 1833 in the winter count constructed by Lone Dog; this spectacular event of 1833 November was so widely noticed that it even inspired an American popular song, *Stars fell on Alabama*! The solar eclipse of 1869 August 7 is another example of an astronomical event noted on a winter count. Petroglyphs by American

FIG. 8. *Leonids on a winter count. Blue Thunder, a Yanktonai Middle Sioux, marked an array of sixteen meteors on his winter count calendar as a mnemonic for 1833, the year of the spectacular Leonid meteor shower. His description of the year is 'The falling-of-stars'. British Museum.*

Indians of comets, meteors and bolides are known. What, then, are the crescent Moon and star representations found by Miller?

Possibly, the drawings represent nothing rarer than a near approach of the crescent Moon to one of the brighter planets (Venus or Jupiter perhaps). Such so-called conjunctions between the Moon and a bright planet can take place several times a year; they are events which have for a long time attracted the attention of astrologers from many cultures because the Moon and planets are supposed to exert influences which augment or conflict at times of conjunction. The crescent Moon and stars is therefore a common symbol; there are more than a dozen representations on present-day national flags – Turkey's, for example.

There would have been many times when such a conjunction occurred during the centuries in which the Arizona sites were occupied. If this recurrent event was of interest, however, it is strange that just a few pictographs have been found in Pueblo Indian sites. On the other hand, the scale of the circle relative to the Moon in both drawings suggests that the star was comparable in brightness to the crescent Moon, and certainly the guest star of 1054 was one of the most notable stars known to have appeared during the time the sites were occupied. Cambridge astronomer Fred Hoyle suggested this to Bill Miller, who then calculated the position of the Moon during July and August 1054. He found that on 1054 July 5 at eight o'clock in the morning (that is, the day after the guest star was noted at its brightest by Chinese astronomers) the crescent Moon would have appeared just above the guest star in the morning twilight, and could have been seen from the Pueblo Indian homes, which had an unobstructed view to the east.

The evidence that the drawings are representations of the new star of 1054 is circumstantial and there are some difficulties – why, for instance, is the crescent Moon reversed in one picture? (Miller suggests that non-astronomically trained people do not worry about this kind of thing and remarks how common it is to find the Moon shown in the wrong orientation in modern illustrations.) It may well be that these drawings really were made by Pueblo Indians 438 years before Columbus sighted land in the Americas, and are indeed records of what amounts to a remarkable pre-Columbian Independence Night fireworks display.

Since Miller's discoveries were announced in 1955, a few more examples have come to light of a crescent-moon-and-bright-star motif, depicted in American Indian art. A third site has been added to the two in Arizona found by Miller and there are five in New Mexico, six in Nevada, five in

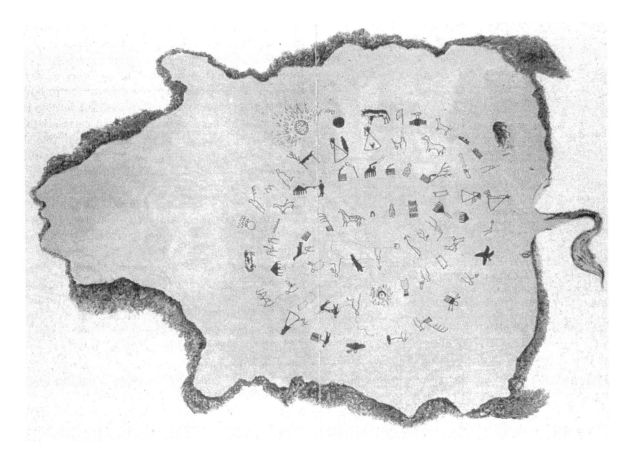

FIG. 9. *Lone Dog's Winter Count. Lone Dog, a Teton Sioux, kept a calendar on a buffalo robe. Each year is tallied with a drawing by which it is named. 1800/ 01 is the year at the centre of the robe. The count spirals out to 1870/71, which is the last drawing near the animal's foreleg. This winter count has three astronomical entries, and they are all explicitly named in the winter counts of other Plains Indians: 1821/22 Star cries out like a buffalo, like thunder, passing along. [A brilliant bolide fell to Earth.] 1833/34 This year is named stars-all- moving year. They feared that Great Spirit had lost control over Creation. [A spectacular display of the Leonid meteor shower.] 1869/70 Morning-sun died. [Solar eclipse of 1869 Aug 7.] Other winter counts describe lunar eclipses and comets, but none survives from the time of the last bright naked eye supernova of 1604. Illustration supplied by Colin Taylor.*

California (including Baja California) and one each in Texas and Utah. A particularly interesting example amongst them is the one in Chaco Canyon, New Mexico, because of its association with other astronomical archaeology. The Moon-and-star motif appears on a sandstone roof. Chaco Canyon is a Sun-watching shrine of the Anasazi Indians (prehistoric Pueblo Indians), with a view of sunrise throughout the year. The site is near to ruins occupied in the mid-eleventh century, and it has been established that the Anasazi observed summer and winter solstices (when the sun rises the furthest north and south respectively) to establish a calendar, presumably for agricultural as well as social purposes. We can imagine the Sun-priest, observing sunrise day by day in order to preserve the calendar, and startled, that morning of 1054 July 5 as the Moon rose an hour or so ahead of the Sun, by the spectacularly bright star off the Moon's southern tip; he may have recorded it immediately on the spot.

The star that appeared for 2 years

What can be learned about the nature of the guest star from the ancient observations? We have, first of all, its approximate position near to the star Zeta Tauri. But more than this, the *Sung Shih*, quoted at the beginning of the chapter, says in another passage about the guest star that it 'remained' in the sky near Zeta Tauri for almost 2 years.

The ancients recognized two kinds of stars, the fixed stars and the wandering stars, or planets. It is now known that, in fact, both the fixed stars and the planets are moving through space at speeds comparable to one another. But the wandering stars – planets – appear to move faster because they are much closer to us than the so-called fixed stars. The planets, in fact, belong to

the nearby solar system; even the closest star is 10 000 times more distant than the farthest planet.

Similarly, comets too speed through the solar system and wheel across the sky, typically in a few months. So, even though the Chinese called some comets 'guest stars' and did not use the term exclusively for new fixed stars, the fact that the guest star was noted to be stationary for 2 years places it certainly as far as the edge of the solar system, beyond the most distant planet, Pluto, and probably outside the solar system altogether.

The Chinese and Japanese records also tell us something of the change in brightness of the guest star. We can deduce from the extracts quoted above that a month before the maximum brilliance of the guest star it was as bright as Jupiter – at its brightest it rivalled Venus, visible during daylight. The guest star ceased to be visible in daylight 23 days after that; it was, finally, fading from the evening sky 630 days later. Because the guest star faded from view to the naked eye into the evening twilight in 1056, it probably was not as faint as the faintest stars visible on the darkest nights.

The evidence that we have of the length of time the star was seen is all consistent with describing this guest star as a supernova, rather than as a nova, since novae fade more quickly. It is now known as the supernova of 1054. It was one of only five supernovae which can have been seen by the unaided human eye within the last 1000 years.

Just as the guest star faded from view, so its memory faded. The vital details of its visibility had been recorded; they were dutifully stored and recopied by Chinese scribes over the centuries, and can now help in the attempt to understand the mysteries of the Crab Nebula, left over from the original momentous explosion.

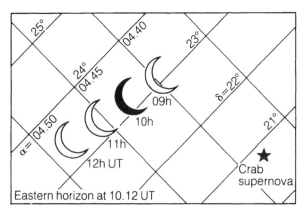

FIG. 10. *Moonrise on 1054 July 5. The Crab supernova and the Moon rose together just after 10 o'clock Universal Time on the morning of 1054 July 5, two and three quarter hours before sunrise as seen from Arizona. For at least two hours the Moon and supernova were visible in the morning twilight. Four successive hourly positions of the Moon are plotted on this diagram, of which the one at 10 h. UT most nearly represents the relative positions of Moon, supernova and horizon at moonrise.*

Sweeping, bushy stars eliminated

What other supernovae were recorded in the oriental observations? The surviving records contain references to a large variety of celestial phenomena. To distinguish records of supernovae from records of comets, aurorae, meteors, lightning and so on, astronomer David H. Clark and historian F. Richard Stephenson have united in an interdisciplinary study. First, they have isolated all mentions of guest stars which had no motion and so excluded moving objects such as comets and meteors. Second, many objects could be eliminated for being described as 'bushy stars' or like a broom ('sweeping stars'). These seem to be references to a comet's tail (though the records are not always unambiguous on this point). This gave them a list of 75 probable novae or

supernovae sighted between 532 BC and AD 1604. Third, Clark and Stephenson picked out the guest stars which lasted a long time, and therefore were not ordinary novae. Of the remainder they recognized the supernovae by the fact that the guest stars were seen in the Milky Way. This is the area of our Galaxy where supernova remnants are now found and therefore where supernovae occur. Clark and Stephenson identified six supernovae which occurred before AD 1500 and left traces in Chinese records. Together with the two supernovae which occurred at the time of the Renaissance, these are the *historical supernovae* (Table 1). Added to the guest star of 1054 there are two supernovae, recorded in oriental observations, whose histories are known in sufficient detail to have enabled radio astronomers to claim with some confidence that they have identified their remnants. They are the supernovae of 1006 and 1181.

The supernova of 1006

Probably the brightest star to have appeared in the sky for the last 1000 years was the supernova of 1006 which blazed from the southern constellation of Lupus. Because it was so far south and was below the horizon to northern European observers, our main sources of information are Arabic, Japanese and Chinese texts. These texts agree fairly well on the position of the phenomenon but it can readily be admitted that they are obscure with respect to other details of its appearance. In spite of the difficulties in interpreting and collating the ancient sources, modern astronomers are convinced that the star was a supernova.

The main eye witness source is the Egyptian Ali b. Ridwan (died in 1061), who lived in the old city of Cairo. He mentions the supernova in an autobiographical footnote to an astrological work

Table 1. *The historical supernovae*

Year (AD)	Constellation	Duration	Maximum magnitude	Remnant
185	Centaurus	20 months	−6?	MSH 14 − 63?
386	Sagittarius	3 months		G11.2 − 0.3?
393	Scorpio	8 months		G348.7 + 0.3?
1006	Lupus	Several years	−9	P 1459 − 41
1054	Taurus	24 months	−5	Crab
1181	Cassiopeia	6 months	+1?	3c58

by Ptolemy, the *Tetrabiblos*, which he was editing. He says that it appeared 'at the beginning of my education' in the 15th degree of Scorpio. The word he uses for the star is *nayzak* which is fairly rare in Arabic astronomical works. Where it does occur, it refers to a very bright comet, which is how later commentators translated the word. Ali, however, says that the object remained stationary relative to the other stars, while the Sun moved into Virgo. It was not, therefore, a comet.

Ali included a list of the exact positions of the planets at the time of his first sighting the supernova. From his list we can be quite sure of the date: 1006 April 30. The modern computation of the planetary positions on this day agrees well with Ali's data. Ali also included the expected astrological conclusions: famine, death and pestilence would break out. Ali detailed his observation of the supernova.

It was a large nayzak, *round in shape and its size two and a half or three times the size of Venus. Its light illuminated the horizon and it twinkled a great deal. It was a little more than a quarter of the brightness of the Moon.*

It is not clear what Ali means when he says that the supernova was about three times the 'size' of Venus. Astronomers then thought that the brighter stars had perceptible discs, this being a physiological effect in the eye. Two Chinese sources even say that it was a half Moon. Ibn al-Jawzi, another Arabic source but from the thirteenth century, says 'it was a large star similar to Venus'. It seems probable that these astronomers, in comparing the supernova to Venus, were trying to give an impression of its brightness compared with the brightest star they knew. Ali, in fact goes further and says it was a quarter of the brightness of the Moon. This may

mean a quarter of the brightness of the Moon when Full, or the brightness of the Moon when only a quarter is illuminated. This latter interpretation is more consistent with the comparison to the size of Venus.

Ali and Ibn al-Jawzi offer the seemingly independent observations that the light from the star 'illuminated the horizon' and that 'its rays on the Earth were like the rays of the Moon'. A Chinese observer noted in the *Sung Shih* that it 'cast shadows'. Another Chinese source says that 'it shone so brightly that objects could be seen by its light'. This all indicates that it was much brighter than Venus.

The most significant European account is from Hepidannus, the monk who wrote the chronicle of St Gall in Switzerland. In the entry for 1006 he says that he saw the star in the extreme south. This places a limit on how far south it could have been since it must have been above the horizon as seen from St Gall (latitude 47½ degrees). Chinese and Japanese observers place it on the present-day border between the constellations Lupus and Centaurus. Its position can thus be tied down to within fine limits.

There are references to the colour of the supernova, but they do not make it very clear. One Chinese observer, the Director of the Astronomical Bureau at the Imperial Court of the Emperor Chen-tung, called it 'a large star, yellow in colour'. This is not to be relied on, however, as an objective description. The reason is partly astrological and partly political. When the supernova appeared, it was so striking that everyone in the Chinese capital, Kaifong, was filled with alarm, and the general opinion was that it was a very bad omen which would be followed by famine and plague. The Director of the Bureau of Astronomy, Chou K'o-ming, was out of town at the time. When he returned, he found the Emperor very anxious and distressed by the situation. Having considered the evidence, he announced that the star belonged to the astronomical category *Chou-po*. This was an excellent move as such a star was an omen of prosperity and could occur in the reign of a wise and just monarch. The Director was later promoted to Librarian and Escort of the Crown Prince. The important characteristic of a *Chou-po* star from our point of view, however, is that it was *always* yellow in colour. The description is therefore inevitable, given the classification, and does not provide us with scientific information. A Japanese report that it seemed to be blue-white may be more objective.

When we ask how long the supernova was visible, we find difficulties, though it was long-lasting. The Chinese *Chu-Su* says that it 'later increased in brightness' but does not say what 'later' means. Ali, however, says that it disappeared suddenly. We are told by several writers that the supernova was visible for 3½ months after which it was too close to the Sun to be seen; this would, however, cause a gradual disappearance, so probably Ali is not referring to this. Venus in the same position would be clearly visible in daylight; therefore we can say that the supernova at that point could not have been brighter than Venus.

After 7 months behind the Sun's glare, the supernova reappeared in the dawn sky between November 24 and December 22. For how long after this the star remained visible is difficult to determine, but it seems to have been more than a year. The Chinese chronicle, the *Sung Shih* refers to a 'Chou-po star' in November 1006 and again on 1016 May 15. No positions are given but if both stars are the same one it must have been erratically visible for up to 10 years. The only indication that both the references are to the 1006

supernova is that these are the only instances of *Chou-po* being used to describe a guest star. The *Sung Shih* gives more evidence that the supernova was visible for several years. In a passage specifically describing the supernova of May 1006, it mentions the first disappearance and reappearance and continues with the significant word 'thereafter' to say that it disappeared near the Sun in the eighth month and reappeared in the eleventh month. Because this suggests a continuing phenomenon, being hidden by the Sun annually, the star must have been visible for at least 2 years.

We may be reaching for too precise a description from those fragmented notes recopied by generations of scribes and summarized by mediaeval historians from the bits and pieces salvaged after the Mongol invasion of China in 1345. Nevertheless, the location is relatively clear and enabled two radio astronomers, Frank Gardner and Doug Milne, to identify in 1965 a radio source at the area where the 1006 supernova appeared. Using the 210 foot radio telescope at Parkes in New South Wales, Australia, they found a structure closely resembling other supernova remnants.

In 1957 Walter Baade had tried to find visible traces of gas where the supernova had appeared, just as he had previously found the nebulae left by other supernovae, but was not able to make any positive identification. The object was too far south for the Californian telescopes. In 1976, however, Sidney van den Bergh (using the Cerro Tololo telescope in Chile, where the constellation of Lupus passes overhead), discovered a faint wispy nebula close to the radio source, ejected from the supernova when Ali b. Ridwan was a student.

The supernova of 1181

Another supernova flared up and briefly amazed observers in AD 1181. The diary of a pessimistic courtier of the Japanese Emperor contains the astrological interpretation that it was

a sign of abnormality, indicating that at any moment we can expect control of the administration to be lost.

The scientific evidence in Chinese and Japanese chronicles is difficult to interpret. Finding out as much as possible is worth some effort, as Richard Stephenson found a supernova remnant called 3C58 in a position which the records indicate was that of the 1181 supernova. We must fit together, as well as we can, the observations of the ancient astrologers and astronomers with the work of the twentieth-century radio astronomers, in order to form a picture of the violent death of the star and its subsequent gradual dissipation into the interstellar matter.

The records present us with the two most common difficulties of this sort of work: one is that they do not agree; the other is that what they do say is imprecise. The star was observed from three geographical areas and we can make some deductions from the three sets of records.

The first to see it were the observers in southern China. The *Sung Shih* tells us that the star was first seen on 1181 August 6. The next to see it seem to have been the Japanese. The *History of Great Japan*, written in 1715, says that a guest star appeared in the north on 1181 August 7. The Chinese in the northern Chin empire reported in the *History of the Chin Dynasty* that they saw the star on August 11.

How long the star was visible is of critical importance in deciding whether or not it could have been a supernova identifiable with the known radio source. The *Sung Shih* says that the star was visible until 1182 February 6 'altogether 185 days; only then was it extinguished'. The account in the *Chin Shih* gives a somewhat shorter time: 156 days. The duration of several months makes it a reasonable assumption that the star was a supernova.

Descriptions in the records of its position are varied: generally they indicate the constellation Wang Liang, the brighter stars of Cassiopeia. The most precise description seems to be the statement in the *Sung Shih*, where the star is said to 'invade' the Chinese asterism Ch'uan-she, a straggle of stars just north of the bright w-shape of the Cassiopeia stars. Noting that the word *fan* was used in the *Sung Shih* to describe some planetary conjunctions where one planet 'invaded' another, Stephenson demonstrated that the word was usually used in cases where the separation was less than a degree. The only known supernova remnant within one degree of Ch'uan-she is called 3C58, and this must be regarded as the probable remnant of the supernova of AD 1181. A now-invisible star marked on a thirteenth-century star map near to 3C58 strengthens this conclusion, but it is not certain.

The evidence for the colour and brightness of this star is largely Japanese. The *Azuma Kagami* tells us:

At the hour hsu [19.00–21.00 local time] a guest star was seen in the northeast. It was like Saturn and its colour was bluish-red and it had rays. There had been no other example since the third year of Kanko [the supernova of AD 1006].

The comparison with Saturn is striking because Saturn would not actually have been visible between those hours, but only towards dawn. Mars would have been visible at the time. Since the Japanese astronomers ignored the ready comparison with Mars and chose instead to mention Saturn, this seems deliberately to imply that the supernova's brightness was closely comparable to that of Saturn. We do not know, however, whether this was the supernova's maximum brightness or whether it was discovered before or after its maximum. Furthermore, this Japanese chronicle says that there had been no other star of its kind since 1006. The supernova of 1006, as we know, was very bright. Perhaps the supernova of 1181 had been very bright too. This interpretation would be consistent with the considerable interest which the star aroused and the number of references to it. But the evidence is not clear enough, and presumably never will be.

The Star of Bethlehem

The best known bright star in history is the Star of Bethlehem. Was it a supernova? To find out, we have to examine the documentary evidence in the same way as we have done for the Chinese and Arabic accounts of other supernovae.

At the birth of Jesus, according to the gospel of St Matthew, chapter 2, 'there came wise men from the east to Jerusalem saying, where is he that is born King of the Jews? For we have seen his star in the east and are come to worship him'. The term translated here as 'in the east' means more precisely 'at its heliacal rising', that is, the wise men saw the star appear in the first rays of dawn.

From Jerusalem, the Magi (who may have been astrologers from Persia or from the Tigris–Euphrates valley civilizations of Assyria, Mesopotamia, or Babylonia), travelled south to Bethlehem following the star which 'went before them, till it came and stood over where the young child was'. This passage is difficult to identify with any astronomical phenomenon since the motion of stars is generally east to west and

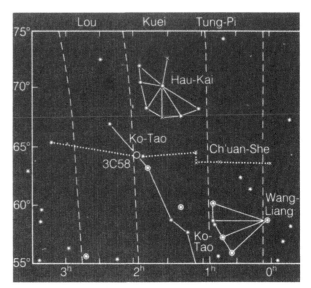

FIG. 11. *Chinese constellations in Cassiopeia. The W of Cassiopeia lies at the bottom of this celestial map of the stars, which are divided into the Chinese constellations. The right-hand half of the W is Wang-Liang, the left-hand is part of Ko-Tao. Hua-Kai is a group of stars centred on 48 Cas. The supernova of AD 1181 was variously reported to be near Hua-Kai and Wang-Liang, and to 'invade' (or, lie within a degree of) Ch'uan-She. The strips of sky labelled at the top of the figure are some of the two dozen slices of sky, each approximately 1 hour of right ascension wide, which correspond to the Chinese 'lunar mansions'. One account places the supernova in lunar mansion Kuei. The supernova remnant 3C58 lies in this lunar mansion, in the constellation Ch'uan-She, and between Kua-Kai and Wang-Liang. It has to be the best candidate for the remnant of the supernova of AD 1181. The diagram is based on research by F. R. Stephenson.*

astronomical objects are so distant that they do not identify one particular terrestrial location but stand equally over whole areas.

Setting this aside for a moment, however, the third item of evidence is from the Proto-evangelium of James (21:2), one of the Apocryphal gospels not included in the Bible, which offers the following:

And he [Herod] questioned the wise men and said to them: 'What sign did you see concerning the new-born King?' And the wise men said: 'We saw how an indescribably great star shone among these stars and dimmed them so they no longer shone, and so we knew that a King was born for Israel'.

Though specific as to the brightness of the star, which would be comparable to the brightness of the Full Moon if it flooded the sky with its light and rendered surrounding stars invisible, this passage raises the difficulty as to how Herod and his advisers could have come to miss noticing the star, unless Herod's question is deliberately disingenuous.

When did the birth of Jesus occur? The first naive attempt at an answer is December 25 in the first year of the Christian era. However, the tradition that Jesus was born at midwinter began about AD 336, possibly because Christians wished to hide their celebrations among the general festival of Saturnalia, or, more likely, because the Church drew the existing pagan festival within the Christian tradition. Luke (2:8) says that at the time of Jesus' birth, shepherds were 'abiding in the fields, keeping watch over their flocks by night'. During winter in Judea flocks were penned, being set free in the spring and guarded by night in the lambing season (March and April). The *fiesta de la Natividad* (to celebrate the birth of Jesus) is held in Spain in the first week of March;

this seems about the right date for Jesus' birthday.

As to the year, the presently accepted calendar of years is from a correlation between Christian tradition and Roman imperial history which is expressed in a calendar reckoned *ab urbe condita* (AUC), from the founding of the city of Rome. The correlation which has been adopted is by Dionysius Exiguus (AD 525) who missed the year 0 between 1 BC and AD 1 and forgot the 4 years during which Emperor Augustus ruled under his own name of Octavian. This would put Jesus' birth in 5 BC or AUC 749. Herod died just before Passover in 4 BC. Jesus was therefore certainly born before then. In his account of Jesus' birth, Luke says that Caesar Augustus had ordered a tax and that this was why Jesus' parents had to travel to Bethlehem. Such an order was issued in 8 BC; the tax would have been collected in the years following. Luke says that tax collection was begun when Quirinius was governor of Syria, but he was not governor until AD 6, although he was an Emperor's legate in Syria between 6 BC and 5 BC, and Luke may have been confused as to his rank. Tertullian, a Roman historian, says that the census at the time of the birth of Jesus was taken by Saturninus, who governed Syria between 9 BC and 6 BC. It seems that the birth of Jesus occurred one springtime between 7 BC and 5 BC.

Having set the date of the Star's appearance, can we determine an astronomical phenomenon for it? Some traditions hold that the Star was a comet; for example, Spanish representations of the Nativity nearly always represent the Star with a tail. Giotto painted the *Adoration of the Magi* in the Scrovegni Chapel in Padua with Halley's Comet as the Star; he had seen the comet during its fourteenth-century apparition but the date on which it appeared nearest the date of the birth of Jesus was the autumn of 12 BC. The date of Jesus' birth rules out this explanation.

A significant astronomical event occurring in 7 BC was a conjunction of Saturn and Jupiter in the constellation Pisces. First suggested as the Star of Bethlehem by Kepler, this was a similar conjunction to the one which led to the discovery of Kepler's supernova. It was not a particularly close conjunction and ones similar to it occur and recur regularly. It hardly justifies the importance which became attached to the Star.

The possibility that the Star of Bethlehem was a supernova (or nova) is one which occurred to astronomers, including Tycho Brahe, after the appearance of the supernova of 1572; this was interpreted by some as signifying a further event of the same kind, possibly the second coming of Christ. What was possibly a nova or supernova was recorded in the *History of the Former Han Dynasty* as occurring in late March or early April in 5 BC, and lasting over 70 days. It appeared in what is now called the constellation of Capricorn. In springtime this constellation rises some 5 hours before the Sun so that the star would have been first observed in the rays of the dawn, as Matthew implies. The relatively short length of time for which the star was visible suggests that it was a nova rather than a supernova. The Chinese records actually call the object a 'broom-star' which is usually used for comets having tails like a brush, although (as pointed out by D. H. Clark, J. Parkinson and F. Stephenson), one record of the well-known supernova of 1572 misclassifies this star in the same way. Whether nova or comet, it was probably not a supernova.

The notion that the Star of Bethlehem was a nova or supernova survives in modern literature in a story (*Nova*) by Arthur C. Clarke. In Clarke's story, a Jesuit astronomer/astronaut discovers the archaeological remains of a beautiful peaceful and cultured civilization, exterminated when their sun exploded in a supernova explosion. His astronomical training leads him to calculate the date of the catastrophe, only to find that it was apparently timed by God to occur so that the light of the explosion reached Earth at just the right moment to proclaim the birth of His Son. Apart from fiction, we might have expected contemporary chronicles of the Chinese or other peoples to have recorded the supernova, if there was one, especially if it was as bright as the Full Moon. No such records have been found. Furthermore, a supernova would not have had the unusual motion attributed to it in Matthew 2:9.

On the other hand, the Star of Bethlehem may have been a genuinely miraculous event for which there is no physical explanation, in which case the above analysis is in vain. It is also possible that the narrative is what in Jewish tradition is called a *midrash*, a historical equivalent of the Christian *sermonizing*. That is, a historical fact presented in a popular manner with decorations adapted to the reader's expected mentality, with echoes of previous parallel events, such as (in this case), the birth of Moses or Abraham. British astronomer David Hughes writes 'no king worth his salt in those days was born without some celestial manifestation. A star greeted the birth of Mithridates (131–63 BC) and Alexander Severus'. This is probably the way to reconcile the difficulties in the various records, which do not have the physical self-consistency of, say, the Chinese accounts of the supernova of 1054, and cannot be taken as compelling. All that is left is the rather vague conclusion that the Star of Bethlehem might perhaps have been the nova of 5 BC but there is no evidence that it was a supernova.

Since bright supernovae do not often explode within sight of the Earth, and a person is unlikely to see one in his life, the impact that they made on those in the ancient world who did witness one may perhaps have been much greater than is implied by surviving records, passed on to us by people who had not witnessed supernovae for themselves. However, two spectacular supernovae appeared during the Renaissance and contemporary accounts of their appearance survive. They appeared at a time when there was an interest in observing and recording natural phenomena for their own sake. The Renaissance supernovae were of great significance to contemporary witnesses and are still to us now.

3

The Renaissance supernovae

Despite the evidence of their own eyes, most people today believe what astronomers tell them about the way the world moves. From an early age we are all taught that the Earth is round, not flat, and that like the other planets it spins on its own axis and orbits the Sun. Yet every day we see the Sun move through the sky, along with the stars and planets. Few have time to prove to themselves that the Earth really does orbit around the stationary Sun.

400 years ago, astronomers assured those who listened that the Earth was stationary and that it was orbited by the Sun. Our ancestors just a dozen generations away believed this. Supernovae helped change their minds.

Crystal spheres and epicycles

For 2000 years, most Europeans accepted the description of the Universe that had been given by the Greek philosopher Aristotle (381–322 BC). Ptolemy of Alexandria turned Aristotle's notion into a quite workable mathematical form in the second century AD. Then in the thirteenth century, Thomas Aquinas took Ptolemy's system as the picture of the Universe to express his theology.

When Aquinas' theology was accepted by the Roman Catholic Church, the cosmologies of Aristotle and Ptolemy helped forge the links of the social order which was seen as a chain stretching from God down to the humblest living creature. A king might hope to be secure on his divinely granted throne because order was seen in the heavens. The philosophy and the science were rarely questioned. Philosophers well understood that if one group of certainties was shattered, all the rest, including the social order, would be subject to doubt.

What was this cosmology that so constrained people's lives? In the Ptolemaic system, the Universe has a set of spheres like an onion.

Completely surrounding the Universe, according to mediaeval philosophy, was the Empyrean where God sat enthroned, accompanied by the souls of the just. Within this lay the *primum mobile*, the First Mover. This was a sphere of an intellectual substance which was thought to be the cause of all movement in the heavens. Movement in the heavens was, in turn, the cause of all movement on Earth.

Inside the *primum mobile* was the eighth sphere, or firmament, the sphere embedded with the fixed stars. (Some saw the necessity for a ninth sphere, the *caelum igneum*, or fiery sphere, to help to account for the apparent movements of the planets.) Below these outer spheres were the spheres of the seven planets, as Saturn, Jupiter, Mars, the Sun, Venus, Mercury and the Moon were then called. The Earth was thought to lie at the very centre of the Universe. Each planet revolved at its own speed about the Earth, the spheres closest to Earth slipping behind the rotation of the firmament by the largest amount.

Aristotle had said that the sphere was the perfect geometrical form. This was why the heavenly bodies were embedded in spheres and why they logically had to move in circles. The lowest sphere contained the Moon and marked the boundary between celestial perfection and the sublunar region of death and corruption which mankind inhabited. With this picture, everything falls into its place. Thomas Aquinas used the idea that we live in a region of death in his doctrine of the fall of Man and his propensity to sin. Although these ideas were interdependent with astronomy, the first changes in beliefs about the solar system had little effect on philosophy. Even though the English philosopher Thomas Digges, like Copernicus, believed that the Sun was the centre of the Universe, he still held to the theory that, above the Moon, all was unchangeable and

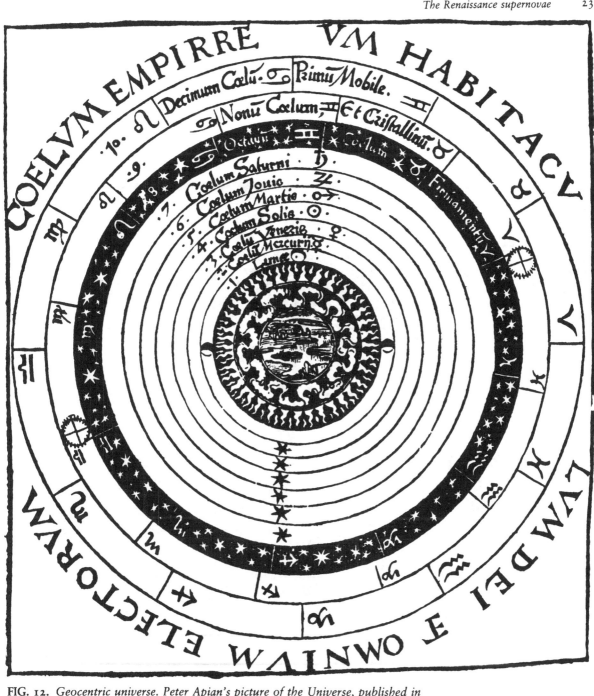

FIG. 12. *Geocentric universe. Peter Apian's picture of the Universe, published in 1524, illustrates the Aristotelian conception of the Universe fused with Christian theology. Outside all is the* Coelum Empirreum, *the highest heaven, home of God and the Elect. Ten spheres are nested within, starting with the* Primum Mobile, *inside which are the ninth sphere of crystal, the eighth sphere containing the Firmament of fixed stars, and the spheres of the seven planets, including number 4, the Sun. Below the sphere of the Moon are the terrestrial regions of fire, the cloudy sky and, at the centre of all, the Earth itself.*

FIG. 13. *The Firmament. Although astronomical historian Owen Gingerich has disproved this picture's historical authenticity, the medieval cosmology of the Universe in which the Earth was at the centre of a sphere of stars is well illustrated in this alleged fifteenth-century woodcut. A pilgrim looks through the Firmament at the mechanisms of the* Primum Mobile *beyond.*

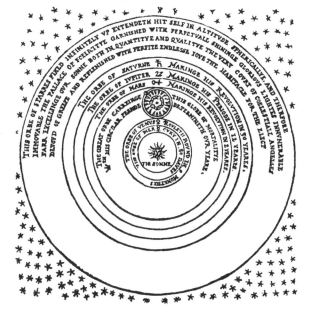

A perfit defcription of the Cœleftiall Orbes
according to the moſt auncient doctrine of the Pytha-
goreans, &c.

FIG. 14. *Heliocentric Universe. Thomas Digges'*
Perfect Description of the Celestial Orbs *was published
in 1576. While trying to come to terms with
Copernicus' idea that the Sun was at the centre of the
Universe, Digges was still a prisoner of the Aristotelian
idea that the stars were immutable. The Orb of stars is
described as 'immovable . . . garnished with perpetual,
shining, glorious lights innumerable', whereas the Orb
of the Earth carries 'this globe of mortality'.*

that death and decay were possible only beneath,
describing in 1576

*the Moon's Orb that environeth and containeth
this dark star [Earth] and other mortal,
changeable, corruptible Elements.*

Shakespeare retained a Ptolemaic Universe in
his plays. The majority of his references to the
stars and planets are to their astrological impact
on human affairs although there are characters
who are made to reject astrological teaching. The
villain, Edmond, in *King Lear*, asserts:

> *I should have been that I am had the
>   maidenliest star
> in the firmament twinkled on my
>   bastardising.*

The majority, however, think like Lorenzo, a
young lover in the *Merchant of Venice*; he looks
up at the stars and tells Jessica:

> *There's not the smallest orb that thou
>   beholdst
> But in his motion like an angel sings
> Still quiring to the young-eye'd cherubim.
> Such harmony is in immortal souls
> But while this muddy vesture of decay
> Doth grossly close us in, we cannot hear
>   it.*

In Lorenzo's world picture, the planets move
around the Earth in their spheres and each one is
inhabited by a spirit, producing a note of great
beauty, so that when all are heard together, the
result is perfection. We who dwell in the sphere
below the Moon are mortal, and cannot hear the
music.

Even in the seventeenth century, Nicolaus
Copernicus' picture of a Sun-centred solar system
was slow to be accepted. John Donne (1571–
1631) thought in terms of crystal spheres:

> *If, as in water stir'd more circles bee
> Produc'd by one, love such additions
>   take,
> Those like so many spheares, but one
>   heaven make,
> For they are all concentrique unto thee.*

At the centre of all the spheres are 'dull sublunary
lovers' whose love could not survive absence
because it was based on the senses alone, and was

therefore imperfect. It contrasted with God's perfect love, from heaven above the highest celestial sphere.

Given the preconception that above the Moon everything was perfect and eternal, the followers of Aristotle accounted for any changes which they saw occurring in the sky by insisting that they took place *below* the Moon. Comets, for example, were said to be atmospheric. According to Aristotle himself, comets were exhalations from the Earth produced by the burning of gases in the atmosphere above the Earth and set on fire by the Sun. Today's view, that they are celestial bodies travelling on elongated orbits which bring them in from the depths of space through the solar system, between the planets, would have been impossible in the Ptolemaic system. The comets would have had to penetrate the crystal spheres in which the planets were embedded, shattering the perfection of circular motions. In Aristotelian science, comets were in fact classed with rainbows, gales, dew, lightning and all other atmospheric phenomena as being meteorological. It was a much later development which restricted the use of the word *meteor* to shooting stars.

The problem of explaining new stars was not very great in the Middle Ages because although oriental astronomers recorded dozens of novae, only a few possible novae were known to European astronomers before 1572. No theory had been developed to explain how such a temporary phenomenon could appear in the external firmament and the novae were explained away. The Star of Bethlehem was thought to be genuinely miraculous and needed no general explanation. The Greek historian Pliny recorded that Hipparchus is said to have observed a new star in 134 BC, and the explanation for this was that what he saw was actually a comet, although Hipparchus should have known the difference

between the point-like appearance of a star and the diffuse appearance of a comet (the word comet derives from *coma*, meaning hair). However, it was firmly held that, whatever he saw was sublunary.

So pleasing was Aristotle's description of the Universe, with its perfect circles and uniform speed of planetary motion, all within the constant firmament, that until the sixteenth century, and even after, most philosophers expended their energy in *saving the appearances*, that is, devising new methods of calculation and minor embellishments of Aristotelian principles which would enable the observations to be reconciled with the theory.

Ptolemy in fact found that the movements of the planets which he observed did not fit the theory of exactly circular orbits. He elaborated Aristotle's theory by claiming that planets moved on *epicycles*, or small circles whose centres themselves moved in circular orbits. In a similar way, Ptolemy dealt with the problem that the observed speeds of the planets were not uniform at all points of their orbits as (according to Aristotle) they should have been, since uniform motion was the perfection expected in a cosmic body. Ptolemy maintained that the speed of a planet was indeed constant, not as seen from the Earth, but as observed from another point in space, the *equant*.

Later philosophers followed in Ptolemy's footsteps, devising elaborations of the epicyclic idea to 'save the appearances'. Even Copernicus who in 1543 took the drastic step of re-ordering the solar system so that the Sun, and not the Earth, was at the centre of the Universe, still presented his new theory of the solar system in terms of epicyclic motion and the result was in many ways more complex than Ptolemy's system.

By about this time it was becoming

increasingly clear that the motions of the planets were hard to explain on a simple epicyclic system. More and more complexities had to be introduced in order to predict accurately the positions of the planets in the sky. But the ancient learning, supported by the Church, was deeply entrenched; as a result, deviants were likely to be prosecuted as heretics. Perhaps it is not surprising that philosophers were prepared to go to great lengths to preserve the old system, which to most people must have seemed the only logical one.

Indeed, when Andrew Osiander came to write the preface to Copernicus' revolutionary book, he had to say that the new theory merely aided calculations and did not necessarily mean that the Sun had ousted the Earth from the centre of the Universe. But none the less the contradiction between theory and observation arose because of the Aristotelian presumption that the Universe above the Moon was perfect. It took the supernova which suddenly exploded in 1572 to shatter the crystal spheres.

Observations of the supernova of 1572

The first recorded observation was made on 1572 November 6 by a Sicilian mathematician, Francesco Maurolyco. He observed a very bright, previously unknown star in the constellation of Cassiopeia. There was great excitement about the new star. 'I am unable to admire enough the new shining of the star of our time' wrote Maurolyco. He noted that he saw the star at the third hour of the night and wrote down its approximate longitude and its angle above the horizon. There is some doubt about whether Maurolyco had seen the supernova before November 6, but a Spanish philosopher, Hieronymus Mugnoz, was teaching an outdoor class in astronomy on November 2 and said afterwards that he would certainly have noticed the new star in Cassiopeia if it had been

visible then. The beginning of the supernova can therefore be placed between 1572 November 2 and 6.

On November 7, Paul Heinzel of Augsburg, Bernhard Lindauer of Winterthur in Switzerland and Michael Mästlin of Tübingen (Kepler's teacher) also saw the supernova. Professor Mästlin satisfied himself that the nova was a star and not a comet. He did this by selecting two pairs of stars in Cassiopeia so that lines between the members of each pair would intersect at the supernova, which was accomplished by holding a thread before his eyes so that it passed through two of the known stars and the new star. In this way, he was able to say that the new star was not moving in relation to the other stars of Cassiopeia. Thomas Digges carried out the same experiment using a six-foot ruler, and placed the star at the intersection of the line joining Beta Cephei to Gamma Cassiopeiae and Iota Cepheri to Delta Cassiopeiae.

Mästlin's and Digges' simple observations were elegant, but the man who won fame and fortune from the supernova of 1572 was Tycho Brahe. Brahe, a Danish astronomer, was an extraordinary man. His personal life was wild and undisciplined in the extreme. He had a gold and silver bridge to his nose necessitated by an injury suffered in a duel when he was a student. Later in his life he lost the private island observatory which had been granted to him by King Frederick because of his arrogant and unjust treatment of his tenants. As a scientist, however, he was meticulous and precise. Indeed, his observations, made before the introduction of the telescope, are recognized as the finest ever made with the naked eye, and achieve an accuracy limited only by the acuity of the eye itself.

Brahe was at the beginning of his career in 1572, and it was in fact the supernova which

FIG. 15. *Tycho Brahe. A Danish nobleman, Brahe was stimulated by his observations of the supernova of 1572 to make accurate observations of the positions of planets at his island observatory at Uraniborg, the City of the Heavens, on Hven. Abandoning this he moved to Prague under the patronage of King Rudolph II of Bohemia and compiled the 'Rudolphine Tables' of planetary motion. In later years he was assisted by his pupil Kepler. From the archives of the Royal Greenwich Observatory.*

inspired him to devote his lifetime to making accurate measurements of the positions of the stars and planets. As Kepler, his pupil, said, 'if that star did nothing else at least it announced and produced a great astronomer'. Brahe's book *De Nova Stella* (1573), in which he first set down his observations and the conclusion that he drew 'about the new star' caused immense interest and

some horror at what were seen as sensational ideas.

In *De Nova Stella*, Brahe described his first sight of the supernova:

Last evening in the month of November, on the 11th day of that month, in the evening, after sunset, when according to my habit, I was contemplating the stars in a clear sky, I noticed that a new and unusual star, surpassing the other stars in brilliancy, was shining almost directly above my head; and since I had, almost from boyhood, known all the stars of the heavens perfectly (there is no great difficulty in attaining that knowledge), it was quite evident to me that there had never before been any star in that place in the sky, even the smallest, to say nothing of a star so conspicuously bright as this.

This cool account is somewhat at odds with Brahe's further admission that, doubtful of the evidence of his eyes, he sought confirmation from his servants and some peasants driving by that they too could see the new star. They could.

The unmoving star

Brahe's most important measurements of the supernova of 1572 were of its position. Although he did not have the advantage of the more accurate instruments which he later acquired for his observatory on the island of Hveen, he did have a large and well-made sextant-type instrument which he had just finished making. He was able to measure the distance of the supernova from the nine principal stars of Cassiopeia, making measurements as accurately as possible with the naked eye. He repeated the measurements at every opportunity, often several times throughout the night. In fact, the star was sufficiently near to the sky's north pole, the star Polaris, so that from Denmark the star never set and could be kept under observation the year round. Like Mästlin and Digges, Brahe found that its position was unchanged for all this time, from hour to hour, from day to day, from month to month to within the accuracy of his sextant. From other measurements of the positions of stars, we know that Brahe's measurements repeated to an accuracy of a few minutes of arc, or less than a tenth of a degree (about the size of a US nickel or a British penny held at a distance of 40 yards). Brahe's measurements firmly put the supernova unmovingly among the other fixed stars.

What motion might Brahe have expected? It was natural for sixteenth-century astronomers to compare the supernova with other transient phenomena, such as comets. Comets move, characteristically, right across the celestial sky in a few months or even weeks. If the new star had been moving at such a rate, Brahe would have detected its motion in a matter of hours, unless, as some of his contemporaries implausibly argued, it was moving directly away from or towards Earth.

He detected no motion over 18 months, eliminating any idea that the new star might have been associated with a planet, since the farthest then known (Saturn) would have moved with motion detectable by Brahe in a week. These general arguments, as well as the observation that the star twinkled just like the other fixed stars, in contrast to the planets which shine without twinkling, were used by Brahe to show that the supernova was indeed a star in the eighth sphere, the firmament. But Brahe had specific arguments to show how distant the star was. His measurements with his sextant showed that the star had no *parallax*, or apparent motion caused by the motion of the Earth, and was certainly beyond the sphere of the Moon. Let us see how he was able to prove this. We will give the

argument in terms of the Earth's rotation, although Brahe would have assumed the Earth to be stationary, and the firmament to be rotating.

Brahe measured the position of the supernova from Heridsvaad when the star was almost overhead at the zenith in the evening sky. 12 hours later, as morning approached, the Earth had rotated halfway round and, since the supernova was circumpolar and could be seen all night, Brahe was able to repeat his measurement.

Brahe would be making the second measurement from a new position in space carried there by the rotation of the Earth. If the supernova were close to the Earth it would no longer be in the same direction. The distance of the supernova is measured in fact by the angle by which it shifts its apparent position, the so-called *parallax*, of the supernova. The smaller the parallax, the more distant the star. Brahe calculated that, at the distance of the Moon, the parallax of the supernova would be about a degree whereas, from his measurements, its parallax could not be in excess of a few arc minutes, making the supernova at least ten times as far away as the Moon.

Such a conclusion forced Brahe to consider that the new star must be above the Moon. He was certain enough to state it in print. In *De Stella Nova* he said:

I conclude therefore that this star is not some kind of comet or a fiery meteor, whether these be generated beneath the Moon or above the Moon, but that it is a star shining in the firmament itself – one that has never been seen before in our time in any age since the beginning of the world.

Astronomers other than Brahe had suspected that the supernova of 1572 had no parallax and must be amongst the fixed stars. Brahe's accurate measurements proved it. In fact, 5 years later he

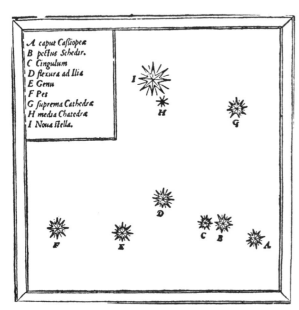

FIG. 16. *Tycho's supernova of 1572. Tycho's map of the 1572 supernova shows the brighter stars of the constellation Cassiopeia with the new star, marked I, brilliantly outshining them. His Latin labels identify the stars by their positions in the mythological Lady in the Chair that the constellation is supposed to represent. Thus A is the head, E the knee and so on.*

was able to show that the comet of 1577 had no parallax either, but travelled across the solar system, passing unhindered through any supposed crystal spheres which might have carried the planets. This again pointed uncompromisingly at change taking place above the Moon. Brahe, however, must have been reluctant to go as far as this. (Perhaps his reluctance was only against printing such conclusions.) In *De Stella Nova* he wrote that all philosophers agree

that in the ethereal region of the celestial world no change of generation or of corruption occurs . . . but celestial bodies always remain the same like unto themselves in every way.

IMAGO CASSIOPEÆ.

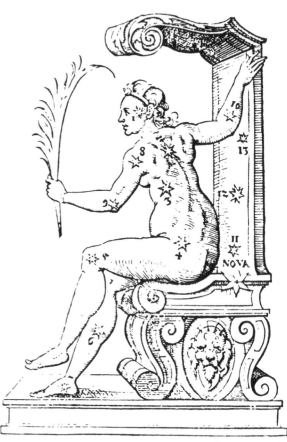

FIG. 17. *Cassiopeia and the supernova of 1572. Czech astronomer Thaddeus Hagecius observed the supernova of 1572 and drew this map of its position ('NOVA'). He had no need to identify the other stars by labels since he added the legendary and decorative figure of Cassiopeia herself to the map.*

His argument was that God had concealed the supernova from earthly eyes since the creation, choosing to reveal it when he wished.

Brahe therefore stood on the brink of a great change in outlook but took refuge in caution. He interpreted his observations of planetary motions in terms of epicycles and of his supernova in terms of suddenly revealed immutability. His pupil, Johannes Kepler, was to overthrow both concepts.

Kepler's supernova of 1604

Tycho Brahe died in 1601, and his work was continued and developed by his former assistant, Johannes Kepler, a German. When the next supernova appeared in 1604, Kepler was working in Prague as court mathematician and tame astrologer for the erratic, and probably mad, Holy Roman Emperor Rudolf II.

In September 1604, many astrologers' eyes were turned towards the region of sky in which Mars and Jupiter were slowly drawing together. The supernova appeared in the nearby constellation of Ophiuchus and, thanks to the conjunction, observers saw the supernova when it first appeared and when its brightness was still increasing. It is rare for a nova of any kind to be discovered before it is at its maximum light, because the increase from obscurity to full brilliance is so rapid.

On October 9, two Italians saw the new star. One was an anonymous physician in Calabria, who reported what he had seen to the astronomer Christopher Clavius in Rome. The other was the astronomer I. Altobelli in Vienna. On October 10, a court official in Prague, J. Brunowsky, caught a glimpse between clouds of the new star and notified Kepler.

Unfortunately, the weather in Prague was cloudy from then until October 17, but some observations were made from places where the skies were clear. When Kepler did see the star on October 17, it was very striking. He wrote that it competed with Jupiter in brilliance and that it was coloured like a diamond. This proved to be near the date of maximum brightness for the supernova as all observers agreed that there was no further increase in brightness after October 15. Kepler made arrangements for continued observations to be made, but in November the supernova was too close to the Sun to be seen. It reappeared from behind the Sun in 1605 January, by which time it was already fading. It continued to be visible until 1605 October, and was carefully observed by Kepler and others until then. When the Ophiuchus region came out from behind the Sun in the spring once again, the supernova had become invisible to the naked eye.

The European astronomers were not alone in observing the supernova. As might be expected, it appears in Chinese and Korean records as well. Apart from being interesting in their own right, the Chinese observations are important because they can be compared with the European results and enable us to assess the accuracy of other Chinese records (such as those of the 1054 supernova). The 1604 star occurs as one of the last entries in a list of guest stars in a document known as the *She-ke*. The entry is as follows:

In the 32nd year of the same epoch, the 9th moon, Yih Chow [1604 October 10] *a star was seen in the degrees of the stellar division* Wei. *It resembled a round ball. Its colour was reddish yellow. It was seen in the southwest until the 10th moon* [1604 October 27–November 26] *when it was no longer visible. In the 12th moon, day* Sin Yew [1605 February 3] *it again appeared in the southeast, in the stellar division* Wei. *The next year in the second moon* [1605 March 24–April 23], *it gradually faded away. In the 8th moon, day* Ting Maou [1605 October 7th] *it disappeared.*

FIG. 18. *Bevis' Cassiopeia. John Bevis' celestial atlas of 1745, in which each page was sponsored by a noble patron, showed the known stars of Cassiopeia, all of which were far outshone by Tycho's supernova, plotted at the edge of the Milky Way. From the archives of the Royal Greenwich Observatory.*

This agrees entirely with the European records and indicates the accuracy of the Chinese observations. Korean records also mention the supernova, describing it as having a greater magnitude than Jupiter, being a reddish-yellow colour and scintillating. All European observers in October 1604 remark upon the red, orange or yellow colour.

Because of the philosophical controversies raging in seventeenth-century Europe, one of the main reasons for the great interest which the supernova aroused was the question of whether or not the star, like Brahe's supernova 30 years before, was among the fixed stars. For this reason, great attention was given to measurement of its position in order to determine whether or not it had parallax. It had none large enough to be measured. Kepler produced a position for it but his figures are less satisfactory than those of David Fabricius at Osteel. From Fabricius's figures, modern astronomers have been able to compute the position to within one minute of arc.

As it was by the Chinese, the supernova was seen by many in Europe to have astrological significance. With his tongue in his cheek, Kepler said that it would bring good fortune to publishers at least, as it would bring a spate of pamphlets and books. He himself at once rushed into print with an eight-page pamphlet in German describing the star and comparing it with Tycho's star of 1572. In 1606 he published his book on the subject, *De Nova Stella*, which was, he said,

a book full of astronomical, physical, metaphysical, meteorological and astrological discussions, glorious and unusual.

As this description makes clear, the book was by no means an astronomical paper as we understand the term, but it does contain astronomy. The Emperor Rudolf II's interest in astrology was such that Kepler was forced to include a great deal of interpretation and prediction. His main advice shows restraint: we should all consider what sins we have committed and pray for forgiveness.

Speculations about the supernova

The burning question to many, however, was this: was the star among the fixed stars and therefore another indication that there was change above the Moon? Kepler's answer was that, like Brahe's supernova, and for exactly the same reasons, the star was indeed above the Moon. Aristotle's model universe had failed again in an important respect, and the rest of it was now under suspicion. Each of its attributes had to be subjected to scrutiny and tested against observation of the real world. Science became not a question of 'saving the appearances', and making small modifications to an agreed-upon philosophy, but of attacking the basis of the philosophy itself.

Kepler went on to use Tycho's observations of the positions of the planets to determine that the orbits of the planets were not perfect circles at all, but were flattened, and formed ellipses. Aristotle's concept of perfection in the regions above the Moon was false in every detail.

Another controversial question concerned the origin of the star. One theory being discussed was that the conjunction of the planets had created such fire that they had ignited it. Kepler did not accept this idea. He believed that there is a material scattered through space which has an inherent ability to gather and then ignite itself. In favour of this idea, he argued that certain simple life forms appear spontaneously, in line with the common belief at the time that maggots appeared spontaneously in dead flesh. Shakespeare's Hamlet refers to this belief: 'For if the sun breed maggots

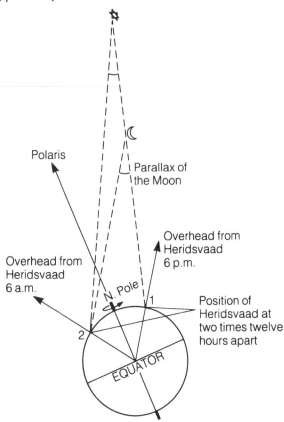

FIG. 19. *Parallax. Brahe's measurement of the parallax of the supernova showed that it was less than the parallax of the Moon and hence that the supernova was more distant.*

FIG. 20. *Johannes Kepler. Kepler's mother was tried for witchcraft and her son derived his income from computing horoscopes. In his early years he published a planetary theory based on the geometric properties of regular solids. He synthesized these mystical notions with Brahe's factual observations of the positions of the planets into Kepler's three laws of planetary motion. In 1604 he observed the most recent bright galactic supernova, and showed the same combination of what we would now call science and pseudoscience by precise and objective measurements of its brightness and by astrological divinations. From the archives of the Royal Greenwich Observatory.*

in a dead dog. . . .' He implies that Ophelia may conceive (though not necessarily spontaneously). Although Hamlet uses the idea ironically, it was a serious belief as Kepler's argument shows.

Controversy over the supernova was not confined to Prague where Kepler was working. In Italy, at Padua, Galileo Galilei was professor of mathematics at the university and had won a reputation for his brilliant lectures and treatises on mathematics and mechanics. Galileo's interest in astronomy, however, did not come to the fore until after 1609 when he made his first telescope. Nevertheless, in 1604 he was acquainted with the problems of astronomy. Tycho Brahe himself had written to Galileo in 1600 and invited him to enter into a scientific correspondence. Galileo seems to have snubbed Brahe and he certainly never entered into extensive correspondence with him. Later, Galileo showed himself extremely hostile to Brahe's ideas, particularly his system of the Universe.

When excitement arose over the 1604 supernova, Galileo, as a leading scientist, was asked to make a statement just as modern Nobel prizewinners are invited by the media to comment on scientific discoveries of which they often know little. Galileo was subjected to some criticism by his sponsors, the Padua city council, because he had not discovered the supernova, to which he somewhat peevishly replied that he had more important things to do than gaze out of the window, on the off chance that he might see something interesting. Galileo was particularly pressed on the question of whether the star was superlunary or merely a meteorological phenomenon as the Aristotelians claimed.

Galileo seems to have been reluctant to make any statement. One reason for his scientific stature was that he required evidence from first hand observations before he would offer interpretations. He expressed his attitude clearly in a parable about a man who understood well the technicalities of how to produce musical notes, but when he held a singing cricket in his hand had no idea how it produced its song. Galileo went on:

The less people know and understand about such matters the more positively do they attempt to reason about them, and on the other hand, the number of things known and understood renders them more cautious in passing judgement about anything new.

When he did give three lectures on the supernova to overflow crowds in the largest hall in Padua, as far as is known he made no definite statement on one side or the other. No texts for the lectures have survived but, from the indirect evidence available, he seems merely to have stated the case that had been so far made by each side, and he declined to draw conclusions. Apparently he began a book on the subject, but he never published it and only a small fraction of the manuscript remains. For the time being, he left the issue in doubt.

Explosions in the minds of men

The philosophical conclusions initiated by the supernovae of 1572 and 1604 and fully confirmed by Kepler's planetary theories, revolutionized the science of astronomy. But the new stars exploded in the minds of non-astronomers as well as in the sky. The first attempts to understand them were astrological. Brahe speculated that a period of peace to be followed by years of violence was indicated by the clear, white light of the star in 1572 since it had been followed by a red, martial light. There were many similar predictions.

Theodore Beza, a French Protestant theologian, wrote a Latin poem suggesting that the star was the Star of Bethlehem, implying that it heralded the Second Coming of Christ. Queen Elizabeth I of England sent for a leading astrologer, Thomas Allen, to ask the meaning of the star, to which 'he gave his opinion very learnedly'. As it had been established early on by Mästlin and Brahe that the star was not a comet, astrologers were being confronted with a phenomenon for which there was no clear European precedent. Sir Thomas Browne considered that the whole business of predictions by astrology had been brought into disrepute:

We need not be appalled by Blazing Stars, and a Comet is no more ground for Astrological presages than a flaming chimney.

From many writers of the seventeenth century, we receive an impression of doubt and uncertainty. Astronomers themselves were disputing the implications of the lack of significant parallax of the supernovae; what was a layman to think? Richard Corbet, Bishop of Oxford, in his letter to Master Ailesbury, written in 1618, expressed this uncertainty:

> *O tell us what to trust to; ere we wax*
> *All stiff and stupid with this Paralax.*
> *Say shall the old Philosophy be true,*
> *Or doth He ride above the Moon think*
>   *you?*

If Corbet sounds unenthusiastic for the new problems of parallax, Henry More, a Cambridge scholar, writing later in 1642 is sceptical and uncomplimentary to those who still held to the old philosophy:

> *That famous star nail'd down in Cassiope*
> *How was it hammer'd in your solid sky?*
> *What pinsers pull'd it out again that we*
> *No longer see it, whither did it fly?*

Probably the most famous expression of discomfort at the new ideas comes from the seventeenth-century poet John Donne. Occasioned by the untimely death of Elizabeth Drury, his poem *An Anatomie of the World* is a consideration of universal decay. In this context, discoveries of new stars point to the end of the old system and are not welcomed for their own sake:

> *And new philosophy calls all in*
> *doubt . . . .*
> *And freely men confess that this world's*
> *spent*
> *When in the planets and the firmament*
> *They seek so many new.*

The certainties of the Middle Ages have been discarded and Man has vacated his place at the centre of the Universe. With this knowledge comes the feeling that Man has lost his former stature. Even the time scale by which we view the Universe is changing. The possibility of greater distances than were dreamed of before brought with it the need for longer time scales:

> *Where is this mankind now? Who lives to*
> *age,*
> *Fit to be made Methusalen, his page?*
> *Alas we scarce live long enough to try*
> *Whether a true-made clock run right or*
> *lie.*

Man himself looks smaller:

> *And as in lasting, so in length is man*
> *Contracted to an inch, who was a spann.*

Repercussions of the new astronomy extended to social order as well as to Man's image of himself. If the Sun is no longer lord of the Universe, since astronomers have had their way with him,

> *They have impaled within a Zodiacke*
> *The free-born Sun,*

what might be the fate of rulers on earth?

> *Tis all in peeces, all cohaerence gone;*
> *All just supply and all relation;*
> *Prince, Subject, Father, Sonne are things*
> *forgot.*

In seventeenth-century England, Donne's fears were more than realized by the social chaos of the Civil War.

Reluctance to abandon certainties in philosophy is understandable. On the other hand there was always some fascination about 'new stars' that could well be felt in an age of explorers. While the map of the Earth was rapidly being drawn larger, there was exhilaration in discovering new depths in the sky. If the size of the world could be doubled, no-one could guess what might happen in astronomy:

> *We have added to the world Virginia and*
> *sent*
> *Two new starres lately to the firmament.*

Donne's reaction to the new small scale of Man against the Universe is here to address him as the author of his own misfortunes and so glorify him into an agent who could actually put new stars in the sky. Consequently there is a note of regret, even blame, when he addresses

> *You which beyond that heaven which*
> *was most high*
> *Have found new spheares and of new*
> *lands can write . . .*

He imagines the discoverers and writers as a controversial group. They come to his mind as an image for disagreement in *A Funerall Elegie*:

> *But as when heaven looks on us with new*
> *eyes*
> *Those new stars every artist exercise,*
> *What place they should assign to them*
> *they doubt,*
> *Argue and agree not till those stars go*
> *out.*

The wording here is clearly a reference to a supernova or nova flaring up and then dying away after a period of months. The frequency of the image in the poetry of a thinking man who was not an astronomer is testimony of the power of the new stars over the mind and imagination.

For many, however, the knowledge of the new stars, dramatic as they were, co-existed peacefully with their view of the Universe and their religion. A man who was able to keep his ideas in separate compartments found no difficulty in appreciating new stars. They were simply more than usually beautiful, perhaps (at the most) portents of good or bad fortune. The poet Edmund Spenser tells a new bride to shine like a supernova:

> *Be thou a new star that to us portends*
> *Ends of much wonder.*

John Dryden in the seventeenth century remembered Tycho Brahe as a means of praising Lord Hastings:

> *Liv'd Tycho now, struck with this Ray*
> *   which shone*
> *More bright i' the Moon than other's*
> *   Beam at Noon*
> *He'd take his Astrolabe and seek out here*
> *What new star twas did gild our*
> *   Hemisphere.*

There are no philosophical questionings, simply an acceptance of the supernova as a beautiful phenomenon which was part of the educated man's horizon at the time.

Tycho himself had compared the observation of the 1572 supernova with the stopping of the sun by Joshua, or the Crucifixion, in its momentous effect on the mind and imagination. Perhaps he would have been pleased by the wealth of writing in which his supernova was celebrated.

The philosophical and cosmological questions which the two stars of 1572 and 1604 had raised, however, were to be answered finally and decisively only after the development of new technology: the telescope.

Answers from the telescope

Historians are still arguing about who invented the telescope. Some say it was a Dutch optician, Hans Lipperhey, around 1608, while others maintain that the instrument had already been known for a decade or more before Lipperhey was granted a license to manufacture telescopes. Similarly, there is considerable evidence that Galileo was not the only person to turn a telescope on the stars in 1609, and that the English astronomer Thomas Harriot had been using telescopes in England at about the same time. What is certain is that it was Galileo who had the power of intellect to understand the implications of what he was seeing, and to relate that to the recent supernovae.

After experiments in producing several prototypes, Galileo designed a satisfactory telescope which gave him a clear view of the sky. Naturally, he pointed his telescope to well-known objects so that he could reveal new aspects of them. He found craters on the Moon, satellites orbiting Jupiter, rings around Saturn, spots on the Sun, new stars in the Pleiades star cluster . . . the list goes on and on. His discoveries revolutionized astronomy.

He recorded his discoveries in *The Starry Messenger* (1610). Between 1623 and 1631 Galileo summarized his cosmology in his *Dialogue concerning Two World Systems*. The book is written as if it were a three-way conversation between an Aristotelian and a Copernican philosopher who try to convince an uncommitted disputant of the truth of their cause. Galileo puts his own view on the Universe through the arguments of the Copernican, Salviati, but he had been forced by the Pope to promise to give equal weight to the official Catholic position as well. He did this through the mouth of a character unsubtly named Simplicius and it is fairly obvious that Galileo's sympathies lay with the other debater, Salviati. One of the characters asks how astronomers could tell whether the new stars were 'very remote'. To this Salviati replies,

Either of two sorts of observations, both very simple, easy and correct, would be enough to assure them of the star being located in the firmament, or at least a long way beyond the Moon. One of these is the equality – or very slight disparity – of its distances from the pole when at its lowest point on the meridian and at its highest [the measurements made by Brahe, and described on p. 29]. The other is that it remained always at the same distance from certain surrounding fixed stars; especially Kappa Cassiopeiae, from which it [the 1572 supernova] was less than one and one half degree distant. From these two things it may unquestionably be deduced that parallax was either entirely lacking or was so small that the most cursory calculation proves the star to have been a great distance from the Earth.

The weight of evidence was now such that, as Galileo says, if Aristotle had been alive he would have changed his mind about the immutability of the heavens. Spots had been seen on the face of the Sun. Comets had been observed which had been

generated and dissolved in parts higher than the Lunar Orb, besides the two new stars, Anno 1572 and Anno 1604, without contradiction much higher than all the planets.

Nevertheless, Galileo had to pay the high

FIG. 21. *Galileo Galilei. Galileo's life spanned the time between the death of Michelangelo and the birth of Newton. His observations covered the sciences of physics, optics and astronomy. He observed the supernova of 1604 and lectured on it, in Italian not Latin, to huge audiences. He wrote of the two Renaissance supernovae in his* Dialogue concerning two world systems, *which effectively dealt a death blow to the Ptolemaic cosmology. From the archives of the Royal Greenwich Observatory.*

price of a summons by the Inquisition and a trial in Rome for stating these views, which were unacceptable because they obliterated the distinction between the corruptible and incorruptible, placing the Earth among the heavens and bringing the heavens down to Earth.

Galileo suffered because his discoveries with his telescope made it impossible to maintain a belief in the unchanging heavens. He found himself obliged to state the conclusions which had in fact been inevitable since 1572. The supernovae had changed European cosmology. What they mean to twentieth-century cosmology is equally momentous but fortunately less likely to arouse animosity or fear.

4

Supernovae in other galaxies

There has been, on average, a bright supernova visible in our Galaxy every 200 years. Unfortunately, not one has been seen and recognized since Kepler's in 1604. Astronomers have had to move gradually towards a better understanding of what supernovae are from observations of supernovae in other galaxies. When extragalactic supernovae were first seen they also caused incredulity and controversy; when the modern founder of supernova research, Fritz Zwicky, began his systematic search for supernovae in other galaxies he was accompanied, he says, 'by the hilarious laughter of most professional astronomers and my colleagues at CalTech'. But before the first supernova was found by deliberate search, the understanding of supernovae continued to grow, helped by various chance discoveries.

A highly remarkable change

On the evening of 1885 August 20, E. Hartwig of the Dorpat Observatory in Russia was discussing the Laplace theory of the origin of the planets with friends. In broad outline his theory, put forward by Pierre-Simon Laplace in 1796, is similar to that widely accepted today.

The starting point of the theory is a huge slowly rotating gas cloud. As the cloud contracts its rotation speeds up until it becomes fast enough to throw off rings of material which then condense into planets. The central part of the cloud eventually becomes the Sun.

Around 1885, when Hartwig's philosophical discussion took place, astronomers were trying to link Laplace's theory with the observational fact that many nebulae were being discovered to have a spiral shape, though none had been resolved into stars. It was not surprising that many people thought them to be planetary systems in formation within our own Galaxy.

As Agnes Clerke, the British astronomer-writer, stated with rash certainty in 1890:

No competent thinker, with the whole of the available evidence before him, can now, it is safe to say, maintain any single nebula to be a star system of coordinate rank within the Milky Way.

Hartwig's friends must have been eager to see one of these mysterious nebulae through a telescope, because he took them out to look at what was then called the Great Nebula in Andromeda, Messier 31. Through Dorpat Observatory's 9-inch refractor very little of interest would then have been visible, since the Moon was approaching Full with its bright light swamping the delicate structure in the nebula. Hartwig had surveyed M31 on three occasions during the previous New Moon period, and must therefore have been astounded to see at the centre of M31 a new star shining brightly where no star had shone before. Clearly, thought Hartwig, this was a central sun appearing as predicted by Laplace's theory.

The star was seen in Hungary by Baroness Podmaniczky on August 22 or 23. She failed to realize its importance. At Heidelberg it was seen by Max Wolf on August 25 and 27 while testing a telescope, but he thought the star was an effect of moonlight. In Rouen, Ludovic Gully at a public night at the newly opened observatory on August 17 had seen the star in the new coudé telescope, but the telescope had been giving trouble in its tests and Gully thought that the appearance of the star was caused by a defect.

Only Hartwig saw the significance of the new star, presumably because he had been studying the region just the week previously. He could not convince the observatory's director of the new star's reality, however, and was not allowed to telegraph the discovery to the central clearing house for astronomical information in Kiel. Hartwig wrote to Kiel anyway but the letters went astray because of the petty theft of the stamps from the envelopes. He was not allowed to announce the 'highly remarkable change in the Great Andromeda Nebula' until he and the director had confirmed the existence of the new star in the moonless sky on August 27. By that time the star had already noticeably faded. Hartwig was able to follow it almost daily as it declined. 180 days after its maximum, which had probably been on August 17, it was beyond the largest telescope's light-gathering power.

The nova was used in renewed attempts to decide whether or not M31 was a galaxy or a nebula. It had become clear that astrophysical arguments were too inconclusive because theories of the behaviour of stars and interstellar gas were not well-enough developed. For example, in 1899 J. Scheiner found that M31 had the spectrum of a collection of sun-like stars rather than the spectrum to be found in hot gaseous bodies. Scheiner correctly inferred that M31 was an aggregate of distant stars. But when V. M. Slipher examined the gas in the Pleiades star cluster, he found that it too had a star-like spectrum, and thought that the Andromeda Nebula might be the same. We now know that the Pleiades gas is dust-laden and simply reflects the light of the bright Pleiades stars.

Clearly only astronomical fact could determine the truth, and the prime fact required was the distance of M31. If M31 were no more distant than the stars of the Milky Way, it could not, in the picturesque phrase of the time, be an 'island universe' of stars.

In the determination of the distance of M31 the nova observed by Hartwig, and now named S Andromedae, played a highly misleading part.

A misleading comparison

In 1911 F. W. Very compared S Andromedae with a nova which occurred in the Milky Way in Perseus in 1901. Nova Persei 1901 brightened in 28 hours from invisibility to naked eye brightness. 3 days later, when it was at its brightest, it was among the half dozen brightest stars. It then faded over a few months back to invisibility.

Within 6 months Max Wolf noticed on photographs of the region of the nova that it had become surrounded by a small nebula. Because this was so soon after the nova outburst, astronomers realized that this nebula could not be gas ejected from the nova. Instead, this must be a reflection nebula. As the burst of light from the star spread at the speed of light into the surrounding space it illuminated a hitherto invisible dark nebula, reflecting light from its successive layers. As the nebula expanded, its brightness diminished: after 2 years it was a dim patch larger than the Moon and it gradually faded away.

Comparing the size of the nebula at various times with the speed of light at which it had been formed, astronomers were able to determine the distance to Nova Persei 1901 as some 500 light years.

At its maximum, Nova Persei was about 250 times brighter than S Andromedae at its maximum. Because the dimming of the light of distant objects depends on their distance squared, it followed that if Nova Persei had been moved about 16 times farther away it would have appeared to be 250 times fainter, the same as S Andromedae. Therefore, if S Andromedae and Nova Persei were alike, the Andromeda Nebula could be no farther than 8000 light years from Earth, not a very large distance and still well within our own Milky Way. From this, F. W.

Very argued that M31 was not another galaxy like our own.

The discovery of further novae in spiral nebulae rekindled interest in this argument in 1917. G. W. Ritchey had photographed a nova in another spiral, catalogue number NGC6946, and this inspired him to re-examine all the photographs of spiral nebulae, including M31, taken by the Mount Wilson 60-inch telescope since 1908. Among the six novae which he found were two in M31 that had passed unnoticed because they were much fainter than S Andromedae. Repeated photography of M31 by Ritchey, Harlow Shapley, John Duncan and R. F. Sanford quickly threw up eight further examples in 2 years, but none nearly as bright as S Andromedae.

It began to be clear that S Andromedae was not a typical nova in M31, and that, in making comparisons between novae in our Milky Way and in the Andromeda spiral nebula, S Andromedae should be ignored. Realizing instantly that the

occurrence of these new stars in spirals must be regarded as having a very definite bearing on the 'island universe' theory of the constitution of the spiral nebulae

H. D. Curtis went on to point out that the difference of brightness between novae occurring in our Milky Way and most of those in M31 was about a factor of 10000. This could be accounted for if the novae in M31 were 100 times the distance of Milky Way novae –

that is, [wrote Curtis] the spirals containing the novae are far outside our stellar system.

Curtis thus found himself a protagonist of the theory that spiral nebulae were galaxies like our own, 'island universes', independent and separated from our Milky Way.

The Great Debate begins

In the same issue of the *Publications of the Astronomical Society of the Pacific* (1917 October) Harlow Shapley developed the same numerical argument as Curtis, but highlighted the problem of S Andromedae. He pointed out that if it was 100 times further than a typical Milky Way nova, then it must have had a luminosity 100 million times that of the Sun, and that

this remarkable result must inevitably follow if spiral nebulae are considered external galactic systems comparable with our own in size and constituency.

During this period Shapley was studying the stars in dense spherical clusters (*globular clusters*), making a stellar census and, by 1919, had come to the conclusion that the study of globular clusters had yielded sufficient knowledge of the luminosity of more than a million stars to show that not one was anywhere near the enormous brightness of S Andromedae. The brightest stars which Shapley found in his census were about 10000 times the brightness of the Sun. Hence, he argued, stellar luminosities of the order 100 million times that of the Sun seemed out of the question, and accordingly the close comparability of spirals containing such novae to our galaxy appeared inadmissable. Shapley found himself on the other side of the debate from Curtis, opposing the 'island universe' hypothesis of the spiral nebulae.

A formal Great Debate on this subject was held on 1920 April 26 at the National Academy of Sciences in Washington, DC. Curtis was forced to grasp the nettle of S Andromedae and to conclude that it seemed certain that the range of brightness of the novae in the spirals, and probably also in our Galaxy, may be very large, as is evidenced by a comparison of S Andromedae with the faint novae found in M31. A division into two classes was not impossible. With this statement Curtis ventured for the first time the concept that, besides ordinary novae, there exists a class of much brighter novae.

The Great Debate itself, of which the argument about the novae was a part, was inconclusive. Neither astronomer convinced the other, nor could their contemporaries decide what was the true status of the spirals. Not until 1923 did Edwin Hubble identify in M31 examples of a type of variable star called cepheids, using the greater power of the 100-inch Mount Wilson telescope to measure the varying brightness of these faint stars. Comparison of these stars with their Milky Way counterparts proved that, according to Hubble's figures, M31 was some 1 million light years away and was indeed a so-called island universe external to our own Galaxy.

Curtis' view of the status of the spirals was shown to be nearest the truth, and the Great Debate has been finally resolved in his favour.

Supernovae revealed

As the evidence began to be assimilated, the startling brightness of S Andromedae became more apparent. Since M31 is a galaxy like our own it contains stars by the billion. S Andromedae at maximum brightness was equal to one sixth of the light from the entire galaxy! Edwin Hubble himself recognized in 1929 this consequence of his observations of the cepheids in M31: S Andromedae belongs to

that mysterious class of exceptional novae which attain luminosities that are respectable fractions of the total luminosity of the system in which they appear.

Up to 1933 haphazard photography of galaxies had thrown up a total of 19 examples of new stars having the property that they nearly equalled the brightness of the galaxies in which they were found. Fritz Zwicky christened the stars *super-novae* in a CalTech lecture course in 1931, and first used the word in public at the 1933 December American Physical Society (APS) meeting; the name lost its hyphen in 1938. In the APS meeting Walter Baade and Fritz Zwicky outlined the fundamental properties of supernovae. To them belongs the credit for being the first to understand the key part that supernovae were to play in modern astronomy. In its entirety the summary of their paper, translated into non-technical language, reads as follows:

Supernovae flare up in every galaxy once in several centuries. The lifetime of a supernova is about 20 days and its brightness at maximum may be as high as 100 million times that of the Sun. Calculations indicate that the total radiation, visible and invisible, is about 10 million times what can be seen. The supernova therefore emits during its life a total energy equal to the amount that the Sun would radiate in a million years. If supernovae are initially quite ordinary stars of mass up to about 10 times that of the Sun, the amount of energy they release is comparable to the energy that would be made if their mass was turned directly to energy. Therefore in the supernova process mass in bulk is annihilated.

In addition, the hypothesis that cosmic rays are produced by supernovae suggests itself. Assuming that in every galaxy one supernova occurs every thousand years the intensity of cosmic rays expected to be observed on Earth is equal to the level actually observed. With all reserve we advance the view that supernovae represent the transitions from ordinary stars into neutron stars which in their final stages consist of extremely closely packed neutrons.

In a brief summary, fully aware of the bizarre nature of what they were saying, Baade and Zwicky mapped out the achievements of the next 50 years' work on supernovae, as the details of their outline were filled in.

In 1934, Zwicky bought a camera just to prove that he and Baade were right. From the top of a building of the California Institute of Technology he repeatedly photographed the rich cluster of galaxies towards the constellation Virgo. He recalled that he was scorned by his colleagues for wasting time. Expecting to find two or three supernovae in 2 years he in fact found none. Though he was almost inclined to give up the systematic search for supernovae because of this setback, Zwicky persuaded George Ellery Hale, Director of the Mount Wilson Observatory, to divert some money from the grant from the Rockefeller Foundation to build the 200-inch telescope in order to construct a new type of camera which would better enable him to find supernovae in other galaxies.

Bernhard Schmidt, an Estonian optical designer, had recently suggested a way of giving an astronomical camera-telescope a much larger field of view, using a corrector lens. Instead of the usual 0.5° or so, the new Schmidt camera photographed a circle of sky 8° across and had an aperture of 18 inches – an ideal instrument for looking at many galaxies at once. Zwicky got his telescope.

In the period from 1936 September to 1939 December, Zwicky and his co-worker J. J. Johnson took 1625 photographs of 175 regions of the sky chosen to contain nearby galaxies. Taking into account the number of galaxies in the field of the telescope the programme amounted to 5150 years continuous observation of an average galaxy. Twelve supernovae were discovered in those 3 years, giving a rate of one discovered

supernova every 430 years per average galaxy. The second which Zwicky discovered, at maximum brightness on 1937 August 22, was more than 100 times brighter than the total light of the galaxy in which it appeared, an irregular spiral galaxy called IC4182, and was the brightest supernova seen so far in this century.

During World War II the search essentially stopped, but Zwicky persuaded Hale to ask the Rockefeller Foundation for nearly $500,000 to construct a larger 48-inch Schmidt telescope, put into operation in 1949. This telescope's first job was to photograph the entire northern sky. The set of photographs which resulted was the Palomar Observatory Sky Survey, which has been a cornerstone of astronomy for 30 years. The Palomar Supernova Search resumed in 1958. Three moonless nights per month were scheduled for the search, which was of 38 fields containing a total of 3003 galaxies in clusters and groups. The photographs taken at night were compared the following afternoon, if possible, with a reference set of photographs of the same fields, so that supernova discoveries could be quickly followed up. The search was officially closed in 1975. The 48-inch telescope had discovered 178 supernovae. Since its inception, the Palomar Supernova Search had discovered 281 supernovae, of which 122 were logged by Zwicky himself. Zwicky noted that each of them cost about $550 to find. For a time the supernova search was carried out by Charles T. Kowal. Now on other tasks, Kowal has informally kept up the search on the two Schmidt telescopes. His personal supernova score passed 100 in 1979. No other observer is in the running for the record of these two men. Supernova searches have been carried out for varying periods at Asiago (Italy), Zimmerwald (Switzerland), Konkoly (Hungary), and Santiago (Chile), using essentially the same techniques as

Zwicky, of taking photographs at different times and looking at them. A *post hoc* supernova search was carried out on the Palomar Observatory Sky Survey when it was realized that the photographs of adjacent fields overlapped but usually had been obtained on different days. When the overlapping edges were searched, supernovae were found, and this is the reason why 1954, the year when most Survey plates were taken, was the bumper year for supernovae (28 discoveries).

Recording the images of thousands of galaxies with a telescope, and comparing them with reference images is a relatively routine business. It is the sort of procedure which, at first sight, lends itself to automation. Indeed, if one could regularly and automatically discover fainter supernovae one could use them as standard candles – lights of a known calibration – in order to measure the distances of thousands of galaxies in the Universe and to determine its scale. Automated search procedures have been proposed in order to make routine discoveries of supernovae, and they have been used experimentally. Stirling Colegate has been developing a computer-controlled telescope with a videcon digital imaging system at the New Mexico Institute of Mining and Technology at Socorro. A Berkeley group plans to use a computer-controlled 36-inch telescope and Charge Coupled Device (CCD) in California. In these programmes the images of galaxies, night by night, are to be compared with digital reference images stored in a computer data file. If a supernova occurs in a galaxy and is detected as an outstanding signal compared to the reference image, then the computer alerts the half-sleeping astronomer (probably a student).

The difficult part of the job is the pattern-recognition logic. Some millions of years of evolution have gone into a human survival strategy based on pattern recognition (spotting deer in the forest, for instance, or snakes in the grass). The result has been that the human eye–brain combination can look at two pictures and almost immediately spot the difference. In looking at pictures of galaxies, a person can automatically compensate for their superficial differences. The two images are not in perfect registry. They were not taken at exactly the same exposure (so that one image is deeper than the other). They were not taken in exactly the same seeing so that one is more blurred than the other. A person can also ignore dust specks. The last 30 years of computer development have not been enough to make machines able to do what a human can do, as quickly, as reliably or as surely. An automated analysis using data from Ed Kibblewhite's plate-measuring machine at the University of Cambridge seemed to recognize 2300 image differences on a pair of Schmidt plates. Half of these were apparent *de*creases in brightness and therefore not supernovae. A quarter of the remainder were near stars, not galaxies, and therefore some problem of scattered light from the stars. Nearly three-quarters of the remainder had quite the wrong shape to be a new star. Between 30 and 60 images were thrown up by the automated search as prospective supernovae. These were looked at by eye. Three-quarters were blemishes on the photographs, immediately recognizable as such to the eye of the student beholder. The handful of possible astronomical objects left over included, on four plates, some variable stars and just three supernovae.

Fully automated searches are still a thing of the future.

SUPERNOVA IN IC 4182

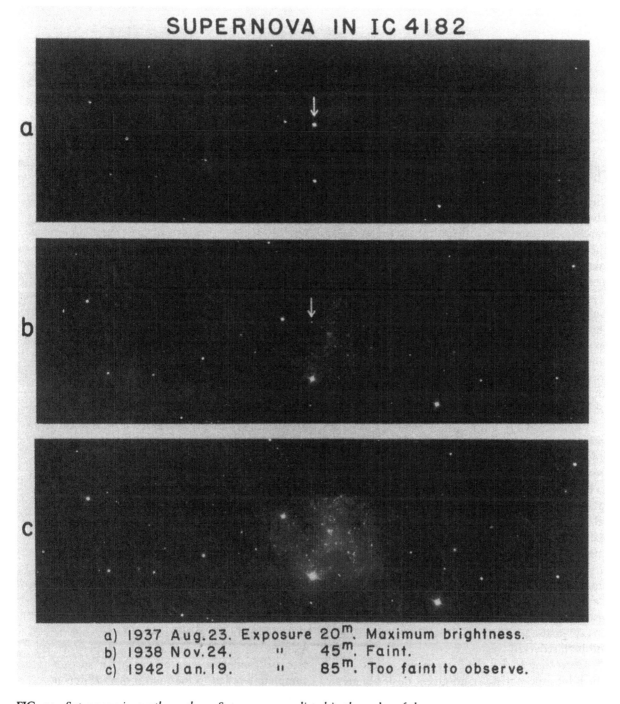

a) 1937 Aug. 23. Exposure 20^m. Maximum brightness.
b) 1938 Nov. 24.   " 45^m. Faint.
c) 1942 Jan. 19.   " 85^m. Too faint to observe.

FIG. 22. *Supernova in another galaxy. Supernovae are listed in the order of the year of their maximum brightness, with a lower case letter to distinguish those occurring in the same year. Supernova 1937c reached maximum brightness (magnitude 8.2) in 1937 August (first picture). Over the next year it faded, becoming invisible by 1942 (last picture). The parent galaxy, IC4182, is a spiral galaxy of total brightness 14.0, so that the supernova at maximum outshone the galaxy by a factor of a 100. Mt Wilson 100-inch telescope photographs from the Hale Observatories.*

The supernova birth rate

The supernova searches give a good idea of the rate at which supernovae are *found* in the galaxies searched. But it is very difficult to determine the true rate at which supernovae actually *occur* in galaxies. There are many systematic effects to be corrected for.

The statistics show that fewer supernovae are found in the fainter galaxies than in the brighter. This is in part because the fainter galaxies are on average further away than the brighter ones, so supernovae in them will be correspondingly less noticeable. It is also because fainter galaxies contain, on average, fewer stars, so supernovae are correspondingly less frequent.

Supernovae are hard to spot when they merge with the bright central regions of galaxies, even though, since there are more stars in the bright areas, you might expect to see more supernovae there. Supernovae occur less frequently in spiral galaxies which are edge-on to us than in spiral galaxies which are face-on. This can hardly be a real effect, since whether we are here to look at a galaxy can hardly have an effect on the evolution of stars in it. The reason is that light from supernovae is progressively dimmed as it traverses the dusty plane of an edge-on galaxy, but passes straight out of the dusty region of a face-on galaxy, making it less likely that the supernovae will be discovered.

Correcting for all these systematic effects, which bias the statistics and give a misleadingly low rate for the occurrence of supernovae, Gustav Tamman has concluded that, in a typical galaxy like ours, the average supernova frequency is as high as one every 20 years, considerably more than Zwicky's first estimate. Up to his death in February 1974, Zwicky himself vigorously disputed such a high frequency for supernovae.

Where have all the supernovae gone?

If supernovae occur as often as once every 20 years in our Galaxy, why don't we see more? Why do we have records of only eight supernovae in the last 2000 years? This would imply a frequency of one every 250 years, completely at odds with Tamman's conclusion. The culprit is the dust and other interstellar material which lies along the galactic plane. This is where supernovae most often occur and it is where the Sun itself lies. From our vantage point near the Sun (on Earth) we see the galactic plane practically edge on, and see the interstellar clouds silhouetted against the massed stars of the Milky Way, slicing through them like the cloud slicing in front of the Moon in the famous image from the surrealistic film *L'Age d'Or* by Luis Buñuel. Look out at the Milky Way on a dark moonless night. In Cygnus and Taurus the Milky Way seems cleft in two. In the Southern Cross a dark hole is silhouetted against the Milky Way – the Coal Sack. All these black patches are relatively near; they are dark clouds of dust hiding the light of stars behind, and they hide the light from supernovae in our Galaxy. Calculations show that only about 40% of all supernovae that occur in the Galaxy can be visible to the naked eye, including supernovae which just become barely visible. Some of these supernovae occur in the daylight sky, and that reduces the possibility that such a supernova is recorded, unless it is so bright that it has not completely faded in the month or two that it takes for the Sun to move on in its journey around the ecliptic. And of course, not all the records of all the supernovae visible in the last 2000 years can have survived.

Even if this explains the lack of historical records of bright supernovae, it is still remarkable that in the 380 years since the invention of the

FIG. 23. *The Milky Way from the constellation of Scutum (top) to Scorpio (bottom). The horizon is blurred as the telescope taking this picture followed the stars. Silhouetted against the massed faint stars of the Milky Way are lanes of dark clouds which blot out the light from stars behind. Galactic supernovae preferentially occur along the centre line of the Milky Way and are heavily obscured by these dark clouds.*

telescope and in the 100 years since comprehensive star charts have been compiled and since astronomers first began regularly to photograph the sky, no modern astronomer has seen a galactic supernova. Even accepting that only about one quarter of galactic supernovae could be discovered, we should still see one every 100 years or so. Why, moreover, are historical supernovae so bunched? There were three in the eleventh and twelfth centuries, a break until two occurred in the Renaissance within 32 years, and none since.

David Clark and Peter Andrews of the Royal Greenwich Observatory and Bob Smith of the University of Sussex set up in 1981 a supernova simulation. They set a computer to choose every 20 years a position at random in our Galaxy for a supernova explosion. Only the flashes from supernovae occurring within 20 000 light years of the Earth were recorded, since this is the distance of a supernova which can just be seen by eye. The light-travel time from the supernova was taken into account when determining the date of the supernova when seen from Earth. The light-travel time is the dominant factor in estimating the rate of discovery of supernovae. Clark emphasizes the point in the case of the two eleventh-century supernovae. The AD 1054 supernova actually exploded 3000 years before the AD 1006 supernova, but was seen 48 years afterwards.

Andrews' programmable hand calculator was set on a 180 000 year simulation, in which it looked at 9000 supernovae, and calculated their date of appearance when seen from Earth. He ran the calculator for several days and the occasional clicks, as it found a visible supernova and logged its statistics on a small printer, added an air of expectancy to the groups of astronomers passing his room. Long before the simulation was finished Andrews knew that the supernovae were bunched

in time because he had noticed how the clicks from the printer came in groups with long intervals between. The simulation showed that the average number of supernovae detected was between 6 and 8 per millennium (1 every 140 years), but the actual number logged in the 180 millennia simulated ranged from 2 to 15, so the fact that we have recorded 4 per millennium is not unusual. The probability that astronomers on Andrews' imaginary Earth in his imaginary galaxy would have to wait over 350 years for a visible galactic supernova was about 7%. The fact that real astronomers have already waited this long since 1604 is therefore also not surprising from a cosmic point of view – just disappointing!

Looking for the next galactic supernova

The next time the Earth is illuminated by the glare of a supernova, the attention of all astronomers will be focussed on it. Whenever the next galactic supernova appears – and it could be tomorrow – the first problem will be to recognize it and to distinguish it from ordinary novae.

Ordinary novae are seen to flare up in our Galaxy at the rate of one or two per year. They are often picked out by watchful amateur astronomers who make a point of searching for them. The sky is so large, and nova searching so frequently fruitless, that professionals from Galileo onwards can find little time for this sort of thing. Amateurs, however, have the time and dedication to learn the appearance of the sky well beyond the limit of naked-eye vision. The game can be rewarding because, as with comets, new stars are sometimes named after their discoverer, who also has the satisfaction of knowing that the world's major telescopes will then be turned towards his star.

As a result of the keenness of the amateur nova patrol, many novae are now being spotted when they are too faint to be seen with the naked eye. This greatly increases the chance that even if a distant, heavily obscured supernova were to appear in the present day, it would still be identified.

Because the occurrence of such things as supernovae cannot be predicted and because they brighten up so quickly, astronomers have now set up an early warning system, run by the International Astronomical Union. Anyone – professional or amateur – discovering an important new object such as a nova or comet notifies the Central Bureau for Astronomical Telegrams in Cambridge, Massachussetts, usually through the facilities of the nearest observatory. The Bureau then cables its subscribers, which include all principal observatories and amateur groups, using a brief and economical code, prefaced by the alerting codenames ASTROGRAM ECHO (for bright novae) or ASTROGRAM FRANCE (for fainter ones).

When an observatory receives one of the innocuous-looking strings of five-digit numbers, there is always a flurry of activity as astronomers decipher the groups and check charts of the right area of sky in their libraries. They then telephone details to colleagues at their observatory's telescopes. It is reasonably safe to predict that every suitable telescope will be pointing at the next galactic supernova within 24 hours of its discovery.

If the next supernova occurs in the direction away from the centre of the Galaxy, then it will almost certainly be bright enough to be visible to the naked eye. Most supernovae, however, are expected to lie towards the galactic centre, just because most of the Galaxy, as seen from the solar system, lies in that direction. If the supernova is relatively close to us, say at a distance of a few thousand light years, it will be

---

Circular No. 2405

CENTRAL BUREAU FOR ASTRONOMICAL TELEGRAMS

INTERNATIONAL ASTRONOMICAL UNION

POSTAL ADDRESS CENTRAL BUREAU FOR ASTRONOMICAL TELEGRAMS.
SMITHSONIAN ASTROPHYSICAL OBSERVATORY. CAMBRIDGE. MASS 02138. USA
Cable Address SATELLITES. NEWYORK - Western Union RAPID SATELLITE CAMBMASS

SUPERNOVA IN NGC 5253

Mr. C. T. Kowal, Department of Astrophysics, California Institute of Technology, telegraphs that he discovered on May 13 a supernova of magnitude 8.5 in NGC 5253 ($\alpha = 13^h37^m.1$, $\delta = -31°24'$, equinox 1950.0). The object, located 56" west and 85" south of the nucleus, was confirmed on May 15. This seems to be the fourth brightest extragalactic supernova ever recorded; the second brightest (= Z Cen), observed in 1895, was also in NGC 5253.

R CORONAE BOREALIS

Recent observations show that this object has been brightening again. C. E. Scovil, Stamford, Connecticut, gives the following magnitude estimates: May 5.17, 11.2; 6.24, 11.3; 7.11, 10.5; 11.28, 9.6; 12.14, 9.6; 13.28, 9.7; 14.18, 9.6. P. Moore, Selsey, Sussex, England, gives: Apr. 29, 12.1; May 7, 10.7; May 10, 10.4.

PERIODIC COMET KEARNS-KWEE (1971c)

In calculating the following ephemeris (cf. IAUC 2330) the ΔT correction (as indicated by several semi-accurate observations by Dr. E. Roemer during July-December 1971) has been applied.

1972/73 ET	α_{1950}	δ_{1950}	Δ	r	m_2
June 12	$2^h40^m.98$	$+22°44'.8$	3.336	2.591	18.3
22	2 59.65	+24 16.9			
July 2	3 18.70	+25 43.7	3.111	2.518	18.0
12	3 38.11	+27 04.3			
22	3 57.77	+28 18.0	2.867	2.451	17.7
Aug. 1	4 17.60	+29 24.2			
11	4 37.46	+30 22.2	2.612	2.391	17.4
21	4 57.16	+31 11.8			
31	5 16.50	+31 53.1	2.353	2.340	17.0
Sept.10	5 35.23	+32 26.3			
20	5 53.01	+32 52.1	2.095	2.297	16.7
30	6 09.54	+33 11.5			
Oct. 10	6 24.39	+33 25.6	1.848	2.264	16.4
20	6 37.14	+33 36.0			
30	6 47.33	+33 43.6	1.622	2.241	16.1
Nov. 9	6 54.48	+33 49.2			
19	6 58.21	+33 52.6	1.434	2.230	15.8

FIG. 24. *IAU Circular. After a brief initial telegram, the world's astronomers are given details of important discoveries by a* Circular *from the International Astronomical Union, like this one, airmailed to thousands of subscribers. This circular, sent out on 1972 May 18, announces Kowal's discovery of the supernova 1972e.*

possible to see the supernova grow in size as it expands into space. Only 40 days after it explodes a supernova at a distance of 3000 light years will appear non-stellar in a telescope. Possibly, if some theories of supernovae are correct, it will be large enough to be seen as a perceptible disc at the time of maximum brightness even when observed by amateur astronomers with moderate-sized telescopes. To the naked eye, the supernova may appear not to twinkle to the same degree as other bright stars, but might glow balefully.

The burst of light from the supernova as it explodes will spread into the surrounding space, at the speed of light, and will illuminate any interstellar material that happens to lie near the supernova (as was the case with Nova Persei 1901). Light from the supernova explosion will be reflected towards the Earth by previously unseen dark clouds of interstellar material and a ripple of light will be seen spreading out through interstellar material, away from the supernova. These light echoes may be visible for hundreds of years, expanding into space until the light is too diffuse to be seen. If they are visible to the naked eye, the light echoes will probably be detectable to a distance of one degree from the supernova, and the light echo will appear as a ring about four times the size of the Moon.

If such a thing is seen surrounding the next galactic supernova, it will provide an easy way to determine its distance. One year after maximum, the radius of the light echo will be exactly one light year. Astronomers will be able to measure its angular distance at this time from the supernova and determine, by trigonometry, the precise distance of the supernova.

Although supernovae emit large quantities of X-rays and gamma rays, this high-energy radiation will be prevented from reaching Earth by the dense expanding material shell ejected by the explosion. There is possibly a brief moment at the beginning of a supernova explosion when there is some hope of detecting a blast of X-radiation, the so-called prompt emission, if there is an X-ray satellite orbiting the Earth at the time. A few years after a supernova explosion, the shell becomes thin enough to be transparent to X-rays again, and they could probably be detected even then if they are strong enough.

As the shell ejected by a supernova disperses into space, it may cool to a point where dust grains can form. These will be warm when they first form and will emit infrared radiation. Therefore, at about the time when X-radiation becomes visible, infrared radiation will also be perceptible while, optically, the supernova fades away.

These ideas about the next galactic supernova are all based on a discussion presented by Sidney van den Bergh in a paper intended to help astronomers to plan observations, should the event occur in their working lifetime. Van den Bergh remarked at the end of his article:

Those who might tend to become discouraged while they wait for this momentous occasion might be slightly consoled by the thought that the light of about 500 galactic supernovae that have already occurred is currently on its way to us!

5

The Crab and its mysteries

While waiting for the next bright supernova to study, astronomers have been studying the remnants of past supernovae, for their own interest and for the light they throw on the supernova phenomenon. Probably more effort has been put into understanding one particular remnant, the Crab Nebula, than any other astronomical object, save the Sun. Solving the mysteries of the Crab Nebula has progressed with the development of new astronomical instruments and techniques, starting with the invention of the telescope itself. Gradually, astronomers have come to understand how significant the tantalizing and infuriating Crab could be in understanding supernovae. No matter how many of its mysteries have been solved, the Crab seems always to hint at another problem.

Discovery of the Crab

Like all the first-discovered nebulae, it was found by chance (not through a systematic search), because the invention of the telescope had posed astronomers a question which most, like Galileo, declined to tackle. How could they systematically survey the whole sky with the telescope and still have time for other studies? Suppose we estimate the field of view of a telescope – the area the astronomer could see at any one time – at 1 degree (twice the diameter of the Moon). This figure is generous for telescopes of the seventeenth and eighteenth centuries. Over the whole sky there are 42 000 such areas of which perhaps one quarter are perpetually below the horizon from European latitudes. Allowing just 4 minutes for the inspection of such an area as the telescope, carried by the Earth's rotation, scans across it, we find that some 2000 hours are needed to sweep the telescope over the whole sky. Astronomical telescopes view the sky at night for about half the time (the other half is spent setting up the telescope, making notes, and so on); it is cloudy in Europe for half to two-thirds of all nights; and half the time the Moon is too bright and floods the sky with light, washing out the fainter stars. Consequently, to be able to observe for 2000 hours in the best conditions requires about 7 years, assuming that the astronomer is dedicated and is prepared to work throughout the year during all hours of darkness. Realistically, such an all-sky survey takes a substantial fraction of a working lifetime. Yet, as telescopes have improved, a few astronomers have repeated examinations of the sky even to the present day.

One such early astronomer was John Bevis (1695–1771). A short account of his life tells us that he studied medicine, astronomy and optics at Oxford, and, although he became a doctor, 'the study of physic afforded him little pleasure in comparison with that of contemplating the celestial bodies and their motions'. He built and furnished an observatory near London and became 'an indefatigable observer'. An active astronomer to the end, he died, aged 76, following a fall while rushing from telescope to clock in the course of observations of the position of the Sun. In about 1745, Bevis compiled the results of his observations of star positions into an atlas, the *Uranographia Britannica*. The costs of preparing the plates for the atlas were so high that the printer went bankrupt and his creditors seized his assets including the engravings of the star charts. A few proofs of the *Uranographia Britannica* had been struck, however, and on these are plotted 16 *nebulae* (Latin for clouds), the term astronomers now use to refer to the unstarlike patches of light which they were beginning to discover.

One set of proofs was shown to the French historian Joseph Lalande on a visit to London. Lalande, in a text on astronomy, records that the great French astronomer Charles Messier had another set. In 1758, Messier, like the whole of the astronomical community, was eagerly awaiting the appearance of Halley's Comet. Many years before, Edmond Halley had realized that this bright comet, then last seen in 1683, returns to the vicinity of the Earth and Sun every 75 years or so. Messier was known as the 'Ferret of Comets' for his assiduous searches for and discoveries of new comets. He actually found a comet near the predicted place, but in fact it turned out not to be Halley's, which arrived later. Messier's new comet passed into the constellation Taurus, and in following it he chanced upon Bevis' nebula. In his own words:

The comet of 1758 being between the horns of the Bull, I discovered on August 28 below the southern horn and a short distance from the star Zeta of that constellation, a whitish light, elongated in the form of a candle flame containing not one star.

He compared the nebula with the comet and said that the nebula was more 'vivid' and more elongated than the comet which seemed 'almost round'.

This discovery was the first of many nebulae found by Messier in the course of sweeping the sky for comets. In 1771 he published a catalogue of all known nebulae with Bevis' nebula in Taurus in first place. After Bevis wrote to Messier pointing out that he had discovered the nebula first, Messier gave him the credit for it. Messier's catalogue of nebulae is known by the initial of its compiler, and therefore Bevis' nebula is called M1.

Although it is first in the catalogue, M1 is by no means the most prominent nebula in the sky. Others, such as the Orion Nebula (M42) are much more spectacular and can easily be seen with the naked eye. While M1 can be seen as a misty patch in fairly small telescopes under good conditions, it needs a moderately large instrument to show it well.

The Crab – unresolved

As telescopes became better and better, astronomers began to find that many of the 103 so-called nebulae in Messier's catalogue were in fact clusters of faint stars packed so closely together that the individual stars could not be separated or *resolved* in poorer telescopes. Naturally, astronomers speculated whether *all* the nebulae would eventually be resolved if large enough telescopes could be trained on them.

For a long time astronomers believed that nebulae were 'cosmical sandheaps too remote to be resolved into stars'. In particular, Bevis' nebula, M1, always gave the impression when seen in better and better telescopes that it was just on the point of being resolved. William Herschel inspected it many times in the course of what he called his 'star-gaging' with the vast telescopes which he had made specially for sweeping the entire heavens. In 1818 he wrote, 'it is resolvable' (but not, notice, 'it is resolved') and went on:

There does not seem to be any milky nebulosity mixed with what I take to be small lucid points. As all the observers agree to call this object resolvable, it is probably a cluster of stars at no very great distance beyond my telescopes' gaging powers.

His son, John Herschel, placed M1 first in a sequence of nebulae which had turned out to be clusters of stars, resolvable by successive degrees, presumably because he too could glimpse the 'lucid points' noted by his father. In fact, M1 was said by John Herschel to be 'hairy' or 'filamentous'.

The nebula was observed by William Parsons,

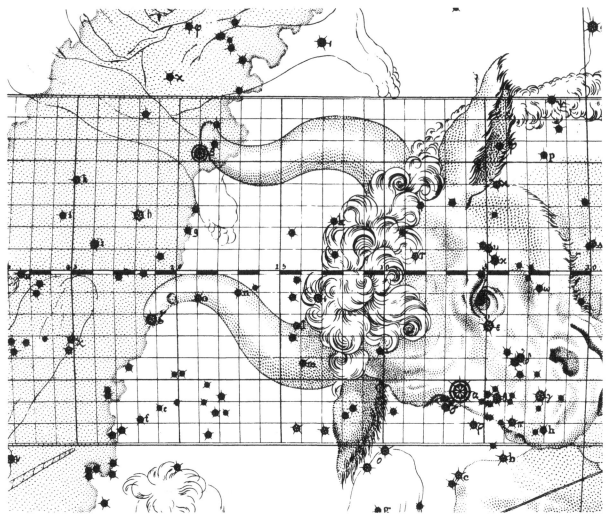

FIG. 25. M1 *in Taurus. Bevis' celestial atlas of 1745,* Uranographia Britannica, *plotted and named the then-known stars of the constellation of Taurus, decorating the chart with a handsome illustration of a Bull with fine, curly hair and intelligent, human eyes. On the chart, Bevis marked the Milky Way and the coordinates of the ecliptic. Near the tip of the Bull's right horn he drew a patch which represented his own discovery of the nebula which later became the first in Messier's list,* M1. *From the archives of the Royal Greenwich Observatory.*

the third Earl of Rosse with his altazimuth telescope of 3-foot aperture. In 1844, writing on his observations of nebulae, Rosse wrote:

Now, as has always been the case, an increase of instrumental power had added to the number of clusters (of stars) at the expense of the nebulae, properly so called; still it would be very unsafe to conclude that such will always be the case.

But, in spite of Rosse's caution here, he too was of the opinion that M1 was a star cluster just beyond the resolution of his telescope as it had been to Herschel's smaller instruments:

It is studded with stars mixed, however, with a nebulosity probably consisting of stars too minute to be recognized.

Rosse added a novel aspect to the description of M1, calling it for the first time the Crab Nebula. He wrote that, with his telescope,

it is transformed to a closely-crowded cluster, with branches streaming off from the oval boundary, like claws, so as to give it an appearance that in a measure justifies the name by which it is distinguished.

Rosse published a curious picture of M1 in 1844 and it has been known as the Crab Nebula ever since.

The Crab figured correctly

Rosse created a memorable name for the nebula, but sadly its actual appearance does not live up to its imaginative title. His contemporary, Dreyer, said that Rosse's 1844 drawing was not at all like the real nebula and, by 1848, Rosse was beginning to regret the drawing which he had published 4 years earlier: he 'would have figured it differently from the drawing in *Phil. Trans.* 1844', he wrote in his observing book on 1848

November 29. He was then observing with his 6-foot telescope, then the world's largest, which he erected near Birr Castle in Ireland. Slung between two towers, the telescope could move only up and down and had very limited ability to track stars as they crossed the north–south line through the telescope. With it, however, Rosse was able to show that many nebulae had a characteristic spiral shape, and we now know that these are distant galaxies mostly too far away for the individual stars to be seen. Although observing with his telescope must have been difficult, the fact that he was able to give such accurate impressions of spiral galaxies shows that what he and his co-workers saw with his larger telescope and what he drew had firm roots in reality.

Rosse measured positions of stars in the Crab Nebula field in 1851, and used them to position features in the nebula accurately during observations made between 1853 and 1855. The drawing was noted as 'finished' on 1855 January 15. The work remained unpublished at his death. His son, Laurence, the fourth Earl of Rosse, continued working in astronomy and made further observations of the Crab Nebula. In 1879 he published an extensive account of the work of the 6-foot telescope from 1848 to 1878, and it is in this work that a lithograph was published from a drawing by R. J. Mitchell, an astronomer on the staff of the Earl of Rosse from 1852 to 1855. It is this representation of the Crab which is close to modern photographs.

The reason for Rosse's more fanciful earlier picture is not clear but others followed him in perceiving something strange about M1. William Lassell, an English amateur astronomer who was a brewer by profession, also remarked on the filaments and 'claws' that he could see when he viewed M1 in 1853 from the clear skies of Malta. Lassell astutely noticed that he could see no more

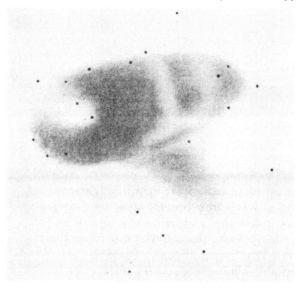

FIG. 26. *The 'Crab'. Lord Rosse's first published picture of M1 christened it with its distinctive name. Rosse later came to regret this representation of the Crab Nebula; it is nothing like his later accurate drawing.*

FIG. 27. *Rosse's Crab Nebula. In 1879 was published R. J. Mitchell's compilation of Rosse's observations of the Crab Nebula and the stars nearby. The accuracy of these observations can be seen by comparing the lithograph with photographs like Fig 30. The lithograph shows the bay to the east (left) and extension to the southwest (lower right) as well as recognizable internal detail. Photo by the Royal Greenwich Observatory from material provided by the Royal Astronomical Society.*

stars in it than were contained in an equal area of other parts of the sky nearby, from which he inferred that the stars he could see within M1 apparently had no connection with it. This marks the beginning of the evidence that the Crab Nebula was not going to be resolved into stars when a suitably large telescope was built, but would remain cloudy and nebular.

The Crab Nebula was first photographed in 1892 by Isaac Roberts, a leading astronomical photographer. Roberts was rather unfairly dismissive about Rosse's 1844 descriptions and Lassell's 1853 drawings, to which he said his photograph showed no resemblance, though he acknowledged that Rosse's later picture had several features that approximately corresponded with the photograph. He described the nebula as

elongated, irregular in outline with a deep bay on one side counterbalanced by a projection on the other. The original negative, he said, showed mottling, rifts, and some star-like condensations in the nebulosity. Apparently Roberts was being careful in saying 'star-like' here; he seems unwilling to say that the condensations *are* stars, and his caution has since been justified.

Stars or gas?

So how does one discover whether the Crab Nebula is made of stars or gas? The first major step forward came when two astronomers at Harvard Observatory, Joseph Winlock and E. C. Pickering, studied the nebula through a device called a *spectrograph*.

The spectrograph ranks next to the telescope

itself in its importance to astronomers, since it has the power to analyse the light from glowing bodies. It is worth spending some time explaining how it does this, and what the results mean.

It was Isaac Newton who first found that white light passed through a triangular glass prism is split into the colours of the rainbow – which scientists called the *spectrum*. In the early spectroscopes, light gathered by an astronomical telescope was split up by prisms and then examined visually with a small telescope. (Modern versions of this device record starlight electronically and produce a picture of the spectrum on television monitors – hence the name spectrograph rather than spectroscope.)

In the first applications of the spectroscope to astronomy it was found that the spectrum of light from most celestial objects was continuous – it spread into *all* the colours of the rainbow. A rainbow is in fact the spectrum of a star – the Sun – spread into its constituent colours by the prism-like action of raindrops. But William Huggins found in 1864 that a large number of nebulae showed not a continuous spectrum, but one in which only a small number of individual colours (called spectral lines) occurred. The explanation for this is that light from a star represents energy given out by the particles of which it is made. A single colour represents a packet of energy of a certain size. If the object on which a spectroscope is focussed is giving out packets of energy of all conceivable sizes, its spectrum contains all colours. This is the case for the light from the dense parts of a star's atmosphere.

A star is composed, like the Earth, of individual particles of matter called atoms which are the basic building blocks of the chemical elements. Unlike atoms on Earth, however, the atoms in a star are hot enough to be moving rapidly back and forth, colliding with each other violently; the force of these collisions is enough to fragment the atoms into some of their constituent pieces. The outer bits of the atoms can be knocked free; these then join in the jostling of the hot atoms. These bits are called electrons. Unless it has been fragmented, each atom has a precise number of electrons, which makes it the kind of atom that it is. The central part of an atom (its nucleus) is positively charged and electrons are negatively charged; therefore the atom is tied together by electrical forces. Only the violent collisions in hot gases at temperatures like those found in stars can overcome the electrical forces. The result is that a star's atmosphere consists of a dense gas of electrical particles, including electrons, rushing back and forth, jostling each other.

Light of any colour (and indeed radio waves and X-rays as well) is commonly produced by just one basic process, though that process can occur in a range of circumstances. Whenever a charged particle is decelerated (or the direction of its motion is changed, which amounts to the same thing), it radiates a pulse of energy, called a *photon*. The wavelength (colour) of the energy depends partly on the violence of the change in motion. So in a star, the random jerks experienced by the electrons produce a whole range of wavelengths, and we see a continuous spectrum.

It is worth mentioning in passing that this continuous spectrum usually has a predominant colour, depending on the temperature of the star. As objects get hotter they glow first red, then yellow, then white, then blue as their peak of energy shifts further to the energetic violet end of the spectrum. The electrons in a hot star generally move faster than in a cool star, and the cool star therefore appears redder than the hot star.

In the same way, the crowd at a football

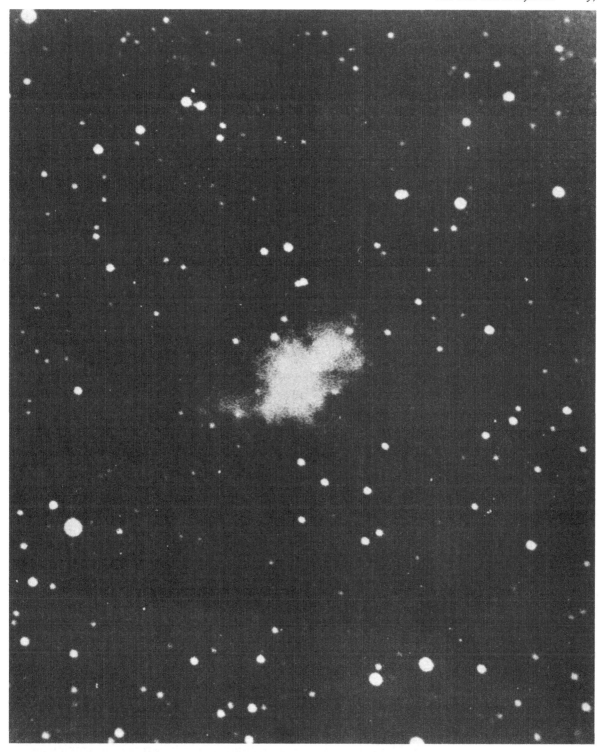

FIG. 28. *Roberts' Crab Nebula. The first photograph of the Crab Nebula, taken by Isaac Roberts, clearly shows the outline of the nebula. Little internal detail is shown on this reproduction, but mottling due to the filaments was reported to be visible on the original negative.*

match have more violent encounters than an equally densely packed crowd at, say, a chess tournament, even though a few individual members of either may be similarly excited or downcast. So there is a spread of energies within both crowds, just as there is a spread of colours emitted by stars of different temperatures, but one degree of excitedness predominates.

Suppose, having looked with a spectroscope at a dense gas such as exists in stars, we turn to inspect a more rarefied gas. This means that there are long distances between atoms, and encounters are few and far between. Little continuous light is given out. The light from such a gas is not the energy emitted from random encounters but from energy changes which take place among the electrons inside the atoms themselves. In a rarefied gas, the atoms have more chance of retaining their own electrons, whereas in the chaotic conditions within a star, the electrons are usually free to move around. All atoms of a given kind contain precisely the same energy levels (like a series of steps) and, as an electron moves from one level to the next, it gives out a packet of energy of the same size as an electron undergoing the same step in any other, similar atom. All atoms undergoing this energy change give out light of the same colour – a spectral line.

Thus a dense material gives out a continuous spectrum but only a rarefied gas can give out spectral lines. Although the full explanation was not known to Huggins, his own observations showed that since obvious star clusters always had continuous spectra, while unmistakably gaseous nebulae always showed bright individual spectral lines, the spectroscope offered a ready means of deciding uncertain cases. Many showed a green spectral line, unknown before its appearance in nebulae, and thought possibly to come from a kind of atom not present on Earth.

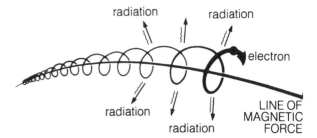

Synchrotron radiation

FIG. 29. *Synchrotron radiation. Following a spiral path around a line of magnetic force, an electron in the Crab Nebula radiates radio, optical and X-radiation by the synchrotron process.*

Indeed, a few years previously, astronomers had found several strong lines in the Sun's spectrum which did not tally with any gas then known on Earth. The mystery gas was christened after *helios*, the Greek for Sun, which is how helium got its name. Similarly, the new element in the nebulae was named *nebulium*. Later analysis showed that the green line was actually the oxygen atom in a state unfamiliar on Earth.

The light from the Crab

Huggins does not seem to have observed the Crab Nebula through his spectroscope but Winlock and Pickering at Harvard in 1868 saw the green nebulium spectral line, which proved that at least part of the Crab Nebula was gaseous.

They reported however that the spectrum showed also an unusually strong continuous component. This must have been puzzling as it suggested that there were more stars than usual embedded in the nebula. In fact, such an interpretation would be wrong. The continuous spectrum of the Crab Nebula is caused in a different way from that of a star – instead of arising by encounters between electrons and other electrons, it arises from the interaction between electrons and a strong magnetic field. This pale clue to a very important process in astrophysics was first noticed in particle accelerators called synchrotrons – its name is *synchrotron radiation*.

Modern colour photographs show that the two spectral components of the Crab Nebula come from distinctly different structures. The nebula consists of a lace of red and green filaments, generally oval in total outline, embedded in and around a tenuous, milky-white light. Many of the filaments are green where the green nebulium spectral line predominates but filaments of a red colour are common. The red spectral line comes from electrons dropping down the first energy step in atoms of hydrogen, and is called H-alpha. It was not seen visually, probably on account of its deep red colour which occurs where the eye's colour sensitivity is poor. Further, weaker spectral lines from other energy steps in the hydrogen atom, called H-beta, H-gamma and so on, are present in the spectrum of the filaments as well, shining green and blue. Other spectral lines from sulphur, helium and neon can be seen with a spectrograph.

Knowing that the nebula has this curious, knotted appearance of filaments embedded in a smoother uncoloured component, we can see now why the eighteenth- and nineteenth-century visual observers always had the impression that the Crab Nebula was about to be resolved into stars. Descriptions such as John Herschel's 'hairy' referred to the filaments, while the knots fooled Rosse into believing that he had really glimpsed stars embedded in nebulosity. If he had called the stars 'lucid points' or 'star-like' (as did William Herschel and Isaac Roberts respectively) he would have given the most accurate visual description of all.

Motions in the Crab Nebula

The mere existence of spectral lines in the spectrum of the Crab Nebula gave astronomers the vital clue that they needed to decide that it was made not of stars but of gas. Measurement of the precise wavelengths of the lines would enable them to study the motions of the gas in the Crab. Nineteenth-century astronomers understood the principle of how this would be accomplished but could not marry theory with practice until the invention of the photographic plate and its application to astronomical spectroscopy. The principle they hoped to exploit was the *Doppler effect*.

Doppler shifts

The Doppler effect is a shift of star's spectral line because of the star's motion towards or away from Earth. It was first explained in 1842 by Christian Doppler, an Austrian physicist, with reference to sound waves rather than to light waves. It was confirmed experimentally by Buys-Ballot who hired a group of musicians who had the gift of absolute pitch so that they could tell precisely what note an instrument was playing without reference to any external standard. Buys-Ballot asked some of the musicians to sit in an open train carriage and play one particular note as the train moved at various speeds along a length of track. While they did this, another musician stood by the track and listened to the note that the musicians were playing. When the train was approaching the stationary listener, he heard a higher note than the one actually being played. When the train was receding, he heard a lower note. Most people, standing at a railway station, have experienced the difference in sound between that of an approaching train and one that is receding.

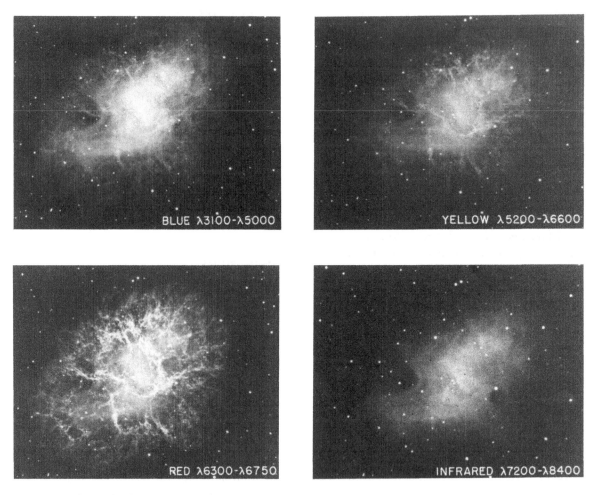

FIG. 30. *Four faces of the Crab. Mt Wilson astronomers filtered the light gathered by the 100-inch telescope in order to restrict the wavelengths used in making these photographs. The wavelength range used to make each is marked on the individual prints. Each shows the relatively featureless glow from the synchrotron process. Some of the wavelength ranges, particularly the red and yellow light, contain strong spectral emissions from the gas in the filaments and the synchrotron image is surrounded and overlaid by the image of the filaments. This photograph is from the Hale Observatories.*

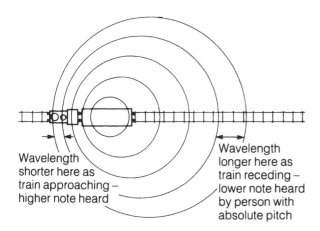

Wavelength shorter here as train approaching – higher note heard

Wavelength longer here as train receding – lower note heard by person with absolute pitch

FIG. 31. *Doppler Effect. In Buys-Ballot's experiment, successive waves of sound were emitted by the musician as he moved along the track on the train. The waves were crowded together (higher note) in the direction of the motion of the train, and spread apart (lower note) to the rear of the train.*

There is close correspondence between what happens in the case of sound waves, and what happens in the case of light waves. We talk about colours or spectral lines instead of musical notes but the two concepts are identical. Instead of a 'higher note', we talk about a 'blue-shifted spectral line', and instead of a 'lower note', a 'red-shifted spectral line'. The degree to which a spectral line is moved towards the blue or red end of the spectrum is a measure of how fast the object which emitted this light is moving towards or away from us. No matter how distant the star, as long as spectral features can be distinguished astronomers can measure its speed along the line of sight. If the star is fairly close, it will also change its position in the sky over a period of time and so its motion in all directions is known.

In practice, the Doppler shift in a typical astronomical object is small. The wavelength of its spectral lines might change, typically, by

0.03%. Astronomers' first attempts to measure Doppler shifts were foiled by their smallness. A spectrum had to be examined in fine detail before they could be perceived. Of course the finer the detail in which you need to look at anything the more light you need to be able to see it, just as to read small print you need a brighter light than is required to read a newspaper headline. The eye could not grasp enough light to 'read' fine Doppler shifts in the first spectroscopes. Only when the astronomer's eye was replaced by a photographic plate to record the spectrum was it possible to measure the Doppler shift caused by motions of stars and nebulae.

The photographic plate, or any other electronic detector, can do what the eye alone cannot: it integrates. That is to say, unlike the eye which perceives a new image 25 times per second, a photographic plate can be put at the focus of a spectroscope for many hours to store up, or integrate, the photons that the sky is sending. Since the spectrum appeared as a permanent picture rather than as a fleeting image in a human eye, the spectroscope was renamed the *spectrograph*. With a spectrograph it became possible to record the wavelengths of light and the way in which light was distributed among the various wavelengths in finer and finer detail.

In 1913, V. M. Slipher turned his spectrograph towards the Crab Nebula and photographed its spectrum. Whereas visual observers, using low-powered visual spectroscopes, had seen individual spectral lines, on his photographic plate he was able to distinguish that every spectral line was doubled: it appeared twice, once shifted a small amount towards the blue end of the spectrum and once to the red. The separation of the red- and blue-shifted lines was larger at the centre of the nebula than at the edges; in fact, a section through the

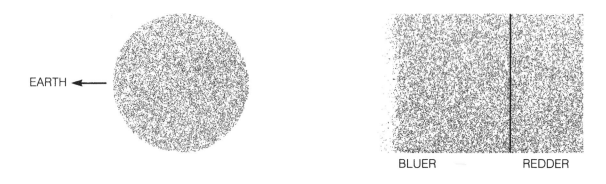

EARTH ←

BLUER — REDDER

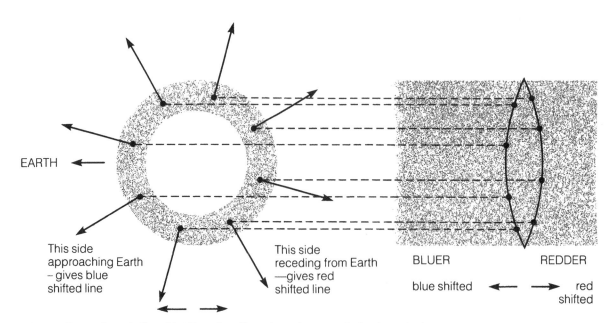

EARTH ←

This side
approaching Earth
– gives blue
shifted line

This side
receding from Earth
—gives red
shifted line

BLUER   REDDER

blue shifted ←   → red shifted

FIG. 32. *Expanding shell and its Doppler effect. A stationary nebula gives single emission lines in its spectrum, but the expanding Crab Nebula shows doubled lines which form a 'velocity ellipse'. Their separation reveals the expansion speed.*

nebula showed elliptical spectral lines. Eventually, Slipher recognized that this was a manifestation of the Doppler effect and that it meant that the light from the Crab Nebula came from two parts, one of which was receding from us and one of which was moving towards us. He correctly deduced that this was because the Crab Nebula was expanding, with the nearer side approaching us and the farther side going away, at a speed up to 1000 km/s.

The Crab in three dimensions

The radial velocity of each portion of the Crab represents the distance of that portion behind or in front of the centre of the nebula. Figure 29 shows how each fragment of the velocity ellipse can be matched to the three-dimensional (3 D) structure of the nebula. At the extremities of the ellipse, for example, the radial velocity of the nebula is zero; these portions of the nebula lie over the nebula's centre – 'in the plane of the sky' is how astronomers put it. At the middle of the velocity ellipse, the radial velocity is at a maximum and these portions of the nebula lie directly in front of (blue-shifted radial velocity) or behind (red-shifted) the centre. If, therefore, an astronomer could determine the spectrum of each part of the nebula, a 3 D image of the Crab would be built up. David Clark and Paul Murdin made 1800 spectra of the Crab Nebula with the Anglo-Australian Telescope in 1978; Andrew Furr and Roger Wood used them to build up a 3 D image in Plate VI which shows that the Crab is like a net, loosely bundled around a hollow ball. The inner surface of the net is the brighter part, the outer surface is fainter. There are filaments which link the two surfaces. Although hollow, the Crab is quite thick: the distance between the surfaces is about equal to the radius of the inner one.

The pale continuous synchrotron light seems held inside the filaments of the net – the electrons which produce the synchrotron radiation are trapped inside the hollow, and so the light which they make comes from the interior of the Crab. It is clear that this is true when you look at the red and blue images of the Crab (Fig. 30). The blue image, which arises principally from the synchrotron radiation, is smaller than the red image, which arises principally from the filaments enclosing the hollow.

Slipher's observations of the Crab Nebula's spectrum were repeated by R. F. Sanford at the Mount Wilson Observatory in 1919. He photographed the Crab's spectrum between the dates of 1918 November 5 and 1919 February 3 patiently building up the image of the spectral lines on the photographic plate – although insensitive by today's standards, Sanford was using one of the fastest emulsions available at that time. 'The result of 48 hrs exposure is disappointingly weak', he wrote. Nevertheless he measured the wavelengths of six of the strongest lines. Then in 1921, C. O. Lampland photographed the Crab Nebula with the 40-inch Lowell reflector and compared that photograph with an earlier one. He was able to see that changes had taken place in the Crab Nebula. This provoked speculation that the changes in appearance were a consequence of the high expansion speeds but Lampland could not distinguish between changes in brightness in individual parts of the nebula and motions caused by expansion, because his pictures had been taken on too small a scale.

Then John Duncan photographed the nebula with the 60-inch reflector at Mount Wilson and compared his photograph with an excellent one made in 1919 by G. W. Ritchey with the same telescope. When Duncan compared the two photographs, he was able to see the changes that

FIG. 33. *Velocity ellipse. In this computer-generated display of an electronically recorded spectrum of the Crab nebula, there are 68 individual spectra from top to bottom. Each plots the intensity of light from a particular spot on the nebula as a function of wavelength. The 68 spots lie across a diameter of the nebula. Each 'hill' on each spectrum represents the fact that a spectral 'line' appears from that spot at a certain wavelength. The 'hills' run from spectrum to spectrum in a correlated way. There are two main spectral lines in this region of the spectrum, the right hand one being brighter than the left. Each spectral line is split by the Doppler effect into an overlapping pair. The whole effect is of two velocity ellipses, whose overall elliptical shape represents the expansion of the Crab. The clumpiness, thickness and irregularity of the ellipse correlates with the clumpiness, thickness and irregularity of the Crab Nebula shell, which is not the idealized smooth, thin, uniform shape in the diagram of Fig. 32. This data was obtained with the Anglo-Australian Telescope by Paul Murdin and David H. Clark.*

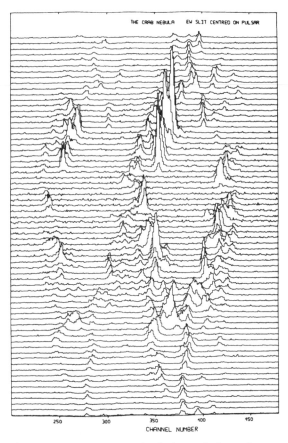

Lampland had announced, and he was able to tell from his larger scale photographs that many of the outer filaments and parts of the nebula had unmistakably moved outwards. Thus the nebula was actually seen to be expanding, and the spectrographic results were clearly confirmed.

Duncan published his result in 1921; in the same year, by a remarkable coincidence, the Swedish astronomer K. Lundmark published a list of novae that had been observed by the Chinese. Number 36 in his list was the supernova of 1054. Lundmark noted that the position of that supernova was very near to M1, the Crab Nebula. In doing so, he seemed to be implying that they were connected and was thus the first to identify a supernova remnant.

Identifying the Crab

The necessary mental jump of connecting the supernova of 1054 with the expansion of the nebula was left to one of the most famous names in astronomy, Edwin Hubble, in 1928. Hubble's name is usually associated with objects outside the Galaxy – he first measured the expansion of the Universe, and the *Hubble Time* defines the age of the Universe itself. He was one of the leading observers of his day, and investigated many nearer astronomical objects, including the Crab. By comparing the size of the Crab with its observed rate of expansion, Hubble was able to estimate

that 800 or 900 years had elapsed since the expansion began.

Hubble was the first to point out the coincidence of position *and age* between the Crab Nebula and the supernova of 1054. He published his inspired guess in a series of popular essays on astronomy and it escaped the attention of professional astronomers. It was not until two Dutch scholars (one an astronomer and one an orientalist) worked on the problem during World War II, that the identification became accepted.

The astronomer was Jan Oort, a leading contributor to knowledge about our Galaxy, and the orientalist was J. J. L. Duyvendak. Working in Holland under the German occupation, they sent their discoveries via Sweden, a neutral country, to N. U. Mayall in the United States. Mayall published the Chinese descriptions of the supernova, which were discovered by Duyvendak, in an astronomical discussion in 1942 – a remarkable example of the way in which science can transcend political divisions, even during wartime.

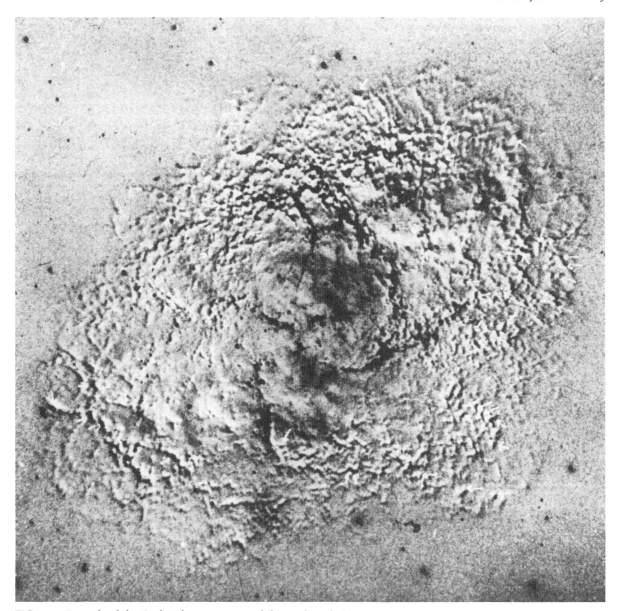

FIG. 34. *Growth of the Crab. The expansion of the Crab Nebula over a period of 14 years is shown graphically in this photograph made by Virginia Trimble. It is a double exposure of a pair of prints taken with the 200-inch Mt Palomar telescope in 1950 and 1964. One, the later print, is a negative with filaments showing black; the other is a positive, with filaments showing white. The stars scattered over the nebula were made to superimpose precisely, so that any movement of the nebula in the intervening years would be clearly seen. Indeed, the black images of the nebula lie outside the white ones, giving a curious 'shadowed' effect which represents the outward expansion of the nebula. Trimble was able to demonstrate, by measuring the relative displacements, so graphically shown, that the filaments were close together at a central point some 800 or 900 years ago, when the Crab supernova exploded.*

FIG. 35. *Crab pulsar and wisps. The central area of the Crab Nebula is shown in this photo taken in 1968 by J. Scargle with the Lick Observatory 120-inch telescope. In this negative print, brighter objects appear blacker. The pulsar is the lower right (southwest) of the two central bright stars. It is touched by a piece of nebulosity known as the Thin Wisp. The bright nebulous area to the upper right is subdivided into two of the three Wisps (the third was not visible at the time that this photograph was taken). Directly opposite the area of the Wisps is the feature known as the Anvil, which was also subdivided into two on this occasion. Apart from the Thin Wisp, the volume around the pulsar seems empty. See the key in the next figure.*

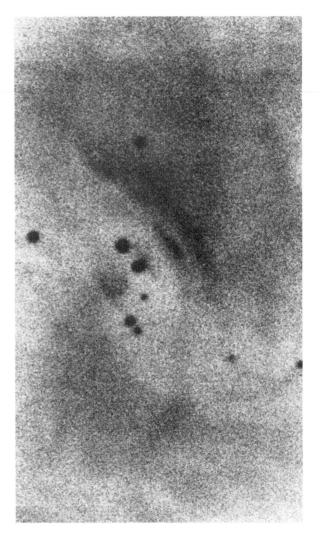

Oort highlighted a slight discrepancy in the records, which has caused a niggling worry ever since. The *Sung Shih* (quoted at the beginning of Chapter 2 on p. 4) quite clearly says that the guest star of 1054 was to the southeast of the star Zeta Tauri. However, the Crab Nebula is to the northwest of Zeta. Astronomers and historians are divided as to the explanation of this discrepancy. The literal minded say that the discrepancy means that the Crab Nebula and the supernova of 1054 cannot be connected. The more imaginative point to the mistakes in other accounts (one of which, for instance, places the supernova in the Pleiades star cluster) and argue that a few discrepant details in some of the records do not outweigh the bulk of the evidence. It is relatively easy to get the orientations of celestial objects wrong, they say. They cite the best evidence of the position of the guest star as a carved stone star map from Suzhou (Soochow). This planisphere was carved on a stele in AD 1247. It was copied from a set of eight charts made by Huang Shang in about AD 1193, themselves representing data compiled in about AD 1100. Approximately 1500 stars are accurately represented on the planisphere, and one is in a position where there is now no star to be seen. It is northwest of Zeta Tauri and, plausibly, represents the guest star whose position was observed 200 years before the stone map was carved, a trace carried in the records that were used to create the map.

There are few astronomers who let the discrepancy in the *Sung Shih* stop them from believing that the supernova of AD 1054 created

the Crab Nebula. The arguments back and forth over the issue illustrate the way that science relies on personal judgement as well as logic. No-one can know everything – in science, as in all human endeavours, understanding comes from assessing the evidence and coming to a conclusion (despite the uncertainties). The scientist who is paralysed with indecision cannot make any advance.

The distance of the Crab

Mayall obtained spectra at the Lick Observatory in California of the Crab Nebula showing very clearly its expansion as evidenced by the Doppler effect. He found that the speed of expansion was

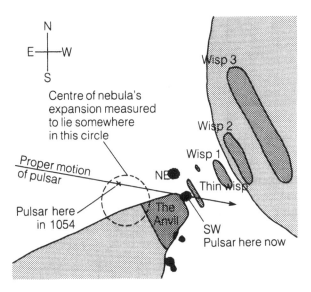

N

E—W

S

Centre of nebula's
expansion measured
to lie somewhere
in this circle

Proper motion
of pulsar

NE

Wisp 3

Wisp 2

Wisp 1

Thin Wisp

Pulsar here
in 1054

The
Anvil

SW
Pulsar here now

FIG. 36. *Map of the centre of the Crab. Waves of activity spread from the southwest star called SW into the Thin Wisp, the Anvil and the three Wisps 1–3. The SW star was at X in 1054 and moves along the line shown, which represents its proper motion. X lies in the circle. The circle encloses the centre of the Crab Nebula's expansion, as measured by Virginia Trimble from the outward motion of the filaments. The coincidence of X with the expansion centre and the spread of the waves of activity from SW were the two facts which pointed at SW as the active star in the Crab Nebula, before it was determined to be the pulsar.*

1300 km/s. Expanding at this speed for 800 or 900 years, the Crab Nebula had become about 7 light years in diameter.

Knowing the true size of the nebula, astronomers could then calculate just how far away it is. Since this method is one often used by astronomers to determine distance, it is worth explaining exactly how it operates.

Consider some object with a fixed size, say a ball 1 foot in diameter. At a distance of, say, 10 feet the ball has a certain apparent size. At twice the distance its apparent size has diminished to half. It appears smaller because of its distance although we know that it still is 1 foot in diameter. Clearly, then, there is a relationship between the distance of an astronomical object, its apparent size and its real size, and if astronomers know any two of these quantities they can calculate the third. In the case of the Crab Nebula, Mayall determined its true size by multiplying the expansion speed by its age, and measured its apparent size in angular degrees, and so could find its distance.

The best modern determination of the distance of the Crab Nebula is 5500 light years, meaning that light takes that length of time to travel from the nebula to Earth. The supernova which the Chinese observed in AD 1054 had actually exploded in around 4500 BC, and the light from the explosion had been travelling for about 5500 years to reach the Earth on 1054 July 4.

Measurements of the expansion of the Crab Nebula tell astronomers not only when the expansion began, but also the place where it began. The filaments had exploded outward and the place from which they radiated is the place where the explosion occurred. The centre of the explosion lies near two stars at the middle of the nebula.

Because of their positions in the sky they became known as the southwest star and the northeast star. The expansion centre is actually now equidistant from both stars. However, the southwest star has a so-called proper motion of its own. (The word 'proper' is not used in any sense of moral rectitude, but because the motion belongs to the star itself and has nothing to do with the motion of the Earth as, say, does its rising and setting.) The northeast star has no discernible proper motion. In their 1942 paper on the Crab Nebula, Walter Baade and Rudolph Minkowski took into account the motion of the southwest star since 1054 and found that *this* star was closer to the original site of the explosion. The northeast star is probably not in the nebula at all, but lies in that direction by chance.

The southwest star lies in a hole in the centre of the Crab Nebula, surrounded by little wisps, bays and filaments which seem to change in brightness and position. The changes were among those first seen by C. O. Lampland in 1921. Walter Baade studied the changes in the centre in detail and found, in 1945, that the brightness changes gave the impression that light was rippling in waves outwards from the southwest star. He took photographs with the 200-inch telescope to prove his point, but had not completed analysing them when he died in 1960. Guido Münch and Jeffrey Scargle subsequently used these plates and later ones which they took themselves to confirm Baade's work.

Baade and Minkowski strongly suspected in their 1942 paper that the southwest star was not only the central remnant of the supernova explosion, but that it was still affecting the central regions of the nebula. They were correct. The southwest star is in fact the Crab Nebula Pulsar, as we shall see in the next chapter, and waves of activity propagate from this star throughout the nebula.

The accelerating Crab

The most accurate estimates of the Crab's rate of expansion actually put the date at which it began to within 10 years of AD 1140, significantly different from 1054. Has there been some strange mistake? Should astronomers be looking for mid-twelfth-century records of a supernova in this part of the sky and for another remnant associated with the guest star of 1054? Historian of science L. P. Williams highlights the fact that astronomers persist in identifying the two objects one with another as an example of the perversity of scientists. In fact, astronomers interpret the very discrepancy as evidence for another of the Crab Nebula's peculiarities, and one which once fooled Baade himself.

The estimate of the date when the expansion began is based on the assumption that the nebula has been expanding throughout its history just exactly as fast as over the last 30 years (the time interval between the photographs used to measure the expansion speed). Astronomers feel that the identification of the Crab with the 1054 supernova is so strong that this assumption has to be wrong. They deduce that the expansion rate has changed. But, rather surprisingly, the nebula is not slowing down: it is speeding up. If it were slowing, its present expansion speed would now be smaller than in the past and we would

mistakenly say that the explosion occurred earlier than 1054. Since it appears from the expansion rate that the explosion occurred 100 years later than 1054, the present speeds are larger on average than the speeds have been over the last 900 years. In fact, the expansion speed is some 400 km/s faster now than originally.

This was first discovered by Walter Baade, using Duncan's measurements, but at first he thought that the observations were not accurate enough. He simply could not believe the result because he would have expected the expansion to slow down as the nebula crashed into the surrounding interstellar material. But it does seem to be the case that the nebula is accelerating. What that implies is that some extra energy is available from somewhere to drive the expansion at faster and faster speeds. The initial energy of the supernova explosion caused it to start expanding, but some additional energy has been pumped into the nebula to make the expansion faster.

Before the mystery of the source of the energy accelerating the filaments could be solved, radio astronomers found further proof that there was an active powerhouse in the Crab, and that it wasn't just using up its inheritance of energy from the supernova explosion itself.

The Crab among the radio stars

Astronomy, like all observational sciences, is full of unexpected discoveries, which pop up through serendipity, as the result of looking for something else. Indeed, the man whose name is now indelibly linked with establishing radio astronomy did not set out to do anything of the sort. The man was an engineer at the Bell Telephone Laboratories, and his name was Karl Jansky.

Bell were interested in the hiss which represented the ultimate limit to the sensitivity of radio reception and transmission, and therefore set Jansky to investigate atmospheric interference. To do this, he built an antenna which he called the Merry-go-Round, because it could be rotated to track down the source of the hiss. In December 1931, Jansky noticed a source of radio interference whose intensity cycled with a period of 23 hours 56 minutes, that is, it was keeping pace with the rotation of the Earth with respect to the stars and not recurring every 24 hours which is the period of rotation of the Earth with respect to the Sun. This led Jansky to conclude that the hiss was of cosmic origin. It peaked in the Milky Way constellation Sagittarius.

Because of commercial pressure, Jansky was unable to continue this investigation, and the first radio maps of the Milky Way were made by an American amateur astronomer, Grote Reber, in the early 1940s, using a home-built backyard antenna. They showed a broad swathe of radio noise coming from the Milky Way, most strongly towards the galactic centre in Sagittarius. There were two subsidiary peaks in the constellations Cygnus and Cassiopeia, blending into the general Milky Way.

Radio astronomy had been studied during World War II, largely because natural cosmic radio noise affected radio reception and the operation of radar. With the ending of the war, it became possible to investigate these phenomena more fully and the opportunity to do so was provided by large quantities of surplus wartime radio and radar equipment. A radio telescope, after all, is no more than a sensitive radio receiver coupled to a directional antenna, such as those used for radar.

A team in England led by John Hey had found during wartime research that the Sun was a source of radio noise when there were large sunspots, and that meteors showed on radar. After the war, this team mapped the intensity of radio waves along the Milky Way and noticed the bright point-like source of radio waves in Cygnus. Hey's radio telescope had a narrower beam than Reber's so that the Cygnus source stood out as more point-like, but the blurriness of Hey's telescope's beam was still 16 times the area of the Moon: its resolution was 2 degrees.

Imagine the optical appearance of the night sky seen with a 2 degree beam. In fact you can simulate such a beam by looking through very defocussed binoculars at the sky. Seen like this, the Moon has no surface detail at all. Its phases cannot be distinguished; all that can be seen is its change of brightness through the month. The appearance of the Milky Way is about the same whether seen through these binoculars or not. The blurred shapes of all but the brightest stars merge into the Milky Way and are indistinguishable from it. The brightest stars, however, can be seen as blurred discs. The view hints at crisper detail beyond the resolution of the binoculars.

The telescope that looked to sea

Arguing by such an analogy, some radio astronomers put effort into determining whether Hey's radio source in Cygnus was truly point-like, and whether there were other radio sources like it. But how could they make a radio telescope that pointed well enough? In Australia, John Bolton and G. J. Stanley used a radio telescope situated on Dover Heights, an eastern suburb of Sydney. They set the telescope on a cliff top overlooking the Pacific Ocean and directed it towards the radio stars as they rose. It received radio radiation not only directly from any radio stars that rose above the horizon, but also from their reflection in the sea. In effect, it functioned in some ways like two connected radio telescopes, the real one 300 feet above sea level on the cliff top, and another one, 300 feet below the sea. Such a radio

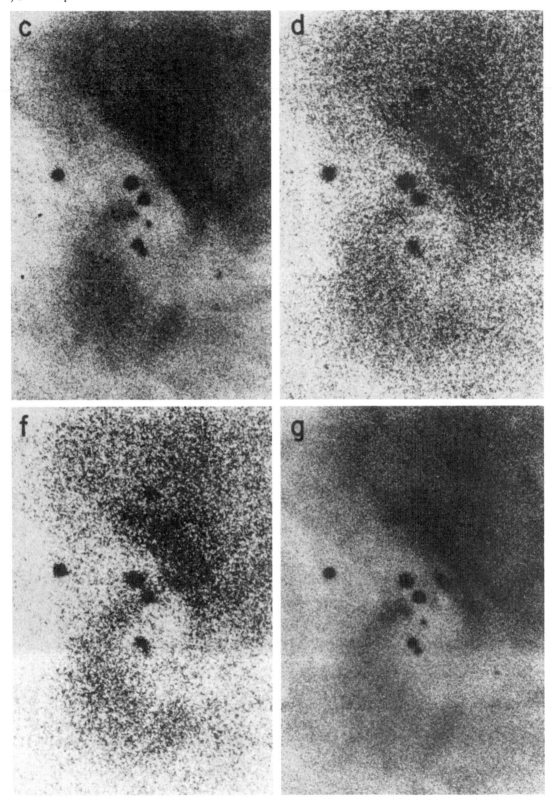

FIG. 37. *Moving wisps. Four photographs of the central area of the Crab Nebula were taken near the time of the abrupt period decrease in 1968 September of the Crab pulsar. The pulsar is the lower right of the central pair. The photographs show different granularity and blurriness due to photographic and atmospheric differences from exposure to exposure. Nevertheless it is possible to see differences in the nebula from exposure to exposure. Photograph c was exposed in September. By the time that the last of the quartet, photograph g, was taken 124 days later, the Thin Wisp had moved from its position immediately adjacent to the pulsar to a position halfway to Wisp 1. Wisp 1 had become a separate patch, distinct from the others and much more compact than in photograph c. These Lick Observatory photos taken by Jeff Scargle confirmed Baade's earlier suspicion that 'waves of activity' spread into the Crab Nebula from the stellar remnant of the supernova of 1054. For the first time, they linked the waves to the release of rotational energy in the Crab pulsar spin-ups.*

telescope, called an interferometer, has the same ability to discriminate fine detail as a single vast radio telescope 600 feet in diameter.

With this telescope, Bolton confirmed the existence of the Cygnus radio source and also saw a radio source in Taurus, which he named Taurus A. His method of determining a radio source's position in the sky consisted in looking at the time at which it rose above the horizon and at the rate at which it ascended above it. However, there was a problem analogous to atmospheric refraction. The effect of air on the light waves from stars at the horizon is to make them visible before they have actually risen above the horizontal. This effect is seen in an exaggerated form in a mirage when light from an oasis below the horizon can be refracted into the gaze of a desert traveller. Normally, atmospheric refraction causes stars and the Sun to rise at least 2 minutes early and set at least 2 minutes late. Another effect is that changes in the degree of refraction cause stars near the horizon to twinkle more than stars overhead. Radio refraction in the ionosphere above Earth is similarly highly variable and is more severe than optical refraction in the atmosphere. Moreover, when radio refraction is strongest, it is least variable and causes least twinkling in the appearance of the radio stars.

Bolton selected the clearest records of Taurus A rising, and in doing so he had selected those with abnormally large radio refraction. Consequently, the first estimate of the position of Taurus A was a long way in error. Compensating for this refraction in later experiments, Bolton took his interferometer to New Zealand, where he observed Taurus A rise from the eastern side of the island, and set from the western side. Any error in time of star-rise was compensated by an equal and opposite error in the time of star-set. Bolton then looked to see what optical sources were in that part of the sky, searching in his only suitable reference work, a star atlas used predominantly by amateur astronomers, *Norton's Star Atlas*. There he found marked M1, the Crab Nebula. Thus in 1948, a radio star was identified with a visible object for the first time.

Bolton and Stanley later used a genuine twin telescope interferometer to observe Taurus A when it was well above the horizon and the effects of refraction were small. They tied the position of Taurus A down to within 15 arc minutes, an area one quarter of that of the Full Moon.

A question immediately arose: was the radio emission coming from the Crab Nebula itself or from a star embedded in the nebula? A two-pronged attack brought the first answers. The initial approach was by various technical advances in making radio interferometers of finer and finer resolution, by placing the radio antennae farther and farther apart. A Manchester group led by Robert Hanbury Brown worked with separations up to 4 km. In Australia, Bernard Mills used two radio telescopes up to 10 km apart to obtain a viewing beam about 1 arc minute wide. The second line of attack was afforded by a lucky chance.

Taurus is not only a Milky Way constellation, it is also on the ecliptic, the yearly path of the Sun, Moon and planets. The Moon, in fact, eclipses the Crab Nebula, and it is possible to see the gradual fading of Taurus A as the Moon moves in front of it. Eclipses in 1956 showed that the radio emission comes from the whole of the nebula, because it is covered gradually as the Moon advances across it.

With the technically superior interferometers now available at Cambridge, Elizabeth Swinbank has made extremely detailed maps of the radio emissions from the Crab, showing that its shape is

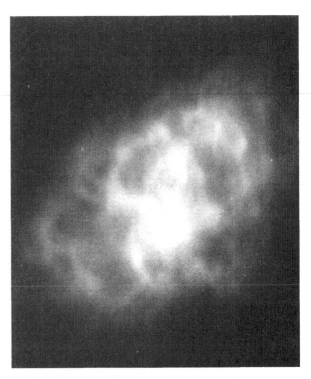

remarkably like that of the white light component of the Crab Nebula, with the same broad hump, dissected by the bays and valleys that show at optical wavelengths. Her radio image of the Crab shows mottling, which agrees in detail with the optical image.

X-rays from the Crab

In 1963, the Crab Nebula was discovered to be emitting X-rays. X-radiation is a form of radiated energy, like light, but a thousand times more energetic. When X-rays shine from outside the Earth onto atoms such as oxygen and nitrogen in the upper atmosphere, the X-rays are quickly and readily absorbed. In this process, electrons orbiting deep within the atoms take up the X-ray energy, causing the electrons to be ejected from the atoms. X-rays travelling at sea level are typically completely absorbed by air after a few hundred feet or so. (Similar X-rays are used in X-ray machines to photograph bones of the body: the areas where X-rays have been absorbed, by just a few inches of flesh and bone, show up as shadows on a photograph.)

Because X-rays are so readily absorbed, the only way astronomers can detect cosmic X-ray sources is by flying X-ray detectors above the atmosphere. Balloons can carry X-ray telescopes above a good deal of the atmosphere but, by its nature, a balloon (needing air to support it) cannot escape completely above the air. Only the most penetrative and energetic X-rays can be detected in this way.

Rockets have the ability to carry instruments completely above the atmosphere in long parabolic trajectories, without going into orbit; it was by this means that the Crab Nebula X-ray source was detected in 1963. The X-ray telescope with which a group of X-ray astronomers at the Naval Research Laboratory (NRL), Washington,

FIG. 38. *Radio Crab. Elizabeth Swinbank and Guy Pooley used the 5 km Cambridge radio telescope of the Mullard Radio Astronomy Observatory to make a radiograph of the Crab Nebula. The nebula is bright in the middle and diffuse. But the filaments also show on the radio picture. The magnetic field bound into the filaments reflects electrons at the filaments and causes a brighter glow of synchrotron radiation there.*

DC, discovered it, accepted X-rays from an area of the sky 20 degrees wide. The year before, a group at the American Science and Engineering Company had flown an experiment which detected an X-ray-emitting region in the Milky Way in Taurus, and the NRL rocket was commanded to sweep over this area. As the NRL X-ray telescope slewed over the Crab Nebula, it recorded a maximum in the number of X-ray counts; the scientists found that the peak was somewhere within 2 degrees of the Crab. In this way, the Crab became the first X-ray source to be identified with a known celestial object (excepting the Sun), just as it had been the first-identified cosmic radio source.

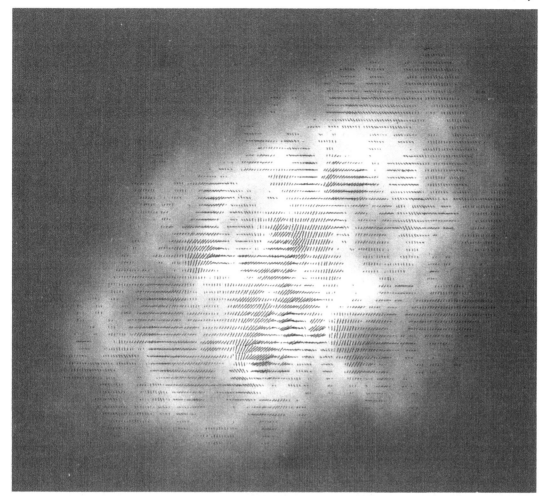

FIG. 39. *Polarization of the Crab. Fritz Zwicky described the polarization pattern of the Crab as a 'basket weave'. The weave is shown as the twists and turns of the radio polarization vectors overlaid on a radiograph of the nebula, in this representation by Elizabeth Swinbank and Guy Pooley, from her thesis. The Crab's magnetic field runs perpendicular to the lines on this diagram. Where they are circumferential around a point on the nebula, the magnetic field radiates from the point. Where the polarization vectors radiate from a point, the magnetic field spirals into the nebula. The polarized radiation comes from electrons gyrating around the magnetic field and that is why there is an intimate relationship between the two.*

The Moon pinpoints X-rays

The NRL group followed up its work with a further rocket, launched so that it would be looking at the Crab during its eclipse by the Moon on 1964 July 7. 3 minutes after launch, the Crab was seen beginning to disappear as the Moon passed in front of it, and it slowly faded away to nothing over the next 3 minutes. During those 3 minutes, the edge of the Moon had scythed across 2 arc minutes of the sky, so that the NRL group had shown that the X-ray-emitting region was approximately the same size as the optical and radio Crab Nebula.

The 'seasons' during which the Moon eclipses the Crab recur every 10 years. A group of X-ray astronomers at the Lawrence Livermore Laboratory of the University of California observed the next lunar eclipse of the Crab in 1974. They were able to tell that the lower-energy X-rays in fact came from an area somewhat smaller than the optical nebula, and centred somewhat west of the star suspected as the centre of the Crab, and near the brightest of the wisps (which Baade and, later, J. D. Scargle had found to shift position and to change in brightness). Another experiment, this time aboard a balloon operated from the Massachusetts Institute of Technology (MIT), showed the very high-energy X-rays to come from an even more compact area centred on the wisps.

These clever experiments with rockets, balloons and the Moon were superceded in the satellite era by direct imaging of the Crab in X-rays. Although far more expensive than a balloon or rocket, a satellite is far less expensive per minute of observing time and allows the build up of X-radiation over a longer period. The succession of X-ray astronomy satellites from Copernicus, Ariel-V and Uhuru in the 1970s culminated in the launch in 1979 of the Einstein Observatory satellite. It contained an Image Proportional Counter (IPC), whose view was a little blurry but which gave spectral information, and a High Resolution Imager (HRI), which gave a sharp picture of resolution 4 arc seconds to images of X-ray sources, including the Crab. The IPC had such a high sensitivity that the Einstein Observatory experimenters would not risk pointing it to the Crab (the brightest X-ray source in the sky) until towards the end of the Einstein mission in March 1981: it was feared that the Crab would burn out the detector. In fact, the IPC did survive. The HRI was exposed to the Crab right at the beginning of the mission, in March 1979, as much to verify the performance of the HRI as to discover new facts about the Crab. The pictures which it obtained confirmed the early experiments. There are no sharp features in the nebula except the pulsar, just a smooth distribution of X-rays centred northwest of Baade's star.

What makes the Crab shine?

The Crab Nebula looks similar in light, radio waves and X-radiation, unlike most other nebulae. And, indeed, all the radio, X-ray and the optical emissions are caused by the same phenomenon, known as synchrotron radiation. When the Russian astrophysicist Iosif Shklovsky proposed this explanation in 1953, the theory was at first thought to be absurd. The model it proposed was of a Crab Nebula full of fast electrons gyrating about strong magnetic fields and radiating their energy over a wide spectrum stretching from radio through optical. Shklovsky predicted that the radiation would be polarized so that when the optical radiation was photographed through a polarizing filter, the nebula would appear streaky along the lines of the magnetic forces which pervade it. This was found to be the case, and the theory of synchrotron radiation was vindicated.

A characteristic of electrons radiating by the synchrotron mechanism is that they lose their energy relatively quickly. The lifetime of a fast-moving electron in the Crab Nebula (that is, the time in which it radiates half its energy) is much less than the 900 year age of the nebula. Therefore, the fast electrons cannot have been spiralling about the nebula since the original supernova explosion; they must have been injected into the nebula since then. The existence of the synchrotron radiation demands that energy must be pumping into the nebula in the form of fast electrons at this moment. Remember that, from the rate of the Crab's expansion, there is other evidence that some powerhouse is still functioning in the Crab. Until the discovery of pulsars in 1968 the source of this extra energy was a mystery.

6

Discovering pulsars

The story of the discovery of pulsars in 1967 is a classic among the many scientific tales of perspiration, inspiration, and just plain luck. Pulsars linked the practical world of the observing radio astronomer with that of the theoretician who, for years, had been talking about mysterious objects called *neutron stars*. And, in explaining how they pulsed and how they emitted radio waves, astronomers found that they naturally provided an explanation for the amazing expansion and acceleration of the Crab Nebula.

The discovery had the added drama that, for a time, it seemed to be evidence for extraterrestrial life, although by the time the news of the first pulsar was published, its discoverers were already quite sure that the signals were not artificial.

Radio scintillation

Even the instrument with which pulsars were first detected has curiosity value in its own right. While most radio telescopes are measured by their size in feet, meters, or even miles, this one has a splendidly archaic name, which never fails to delight non-English astronomers unfamiliar with this unit of area: it is the 4½ Acre Telescope at Cambridge, England.

It was designed for a purpose which had nothing to do with the regions of space where pulsars lurk: to pick out those enigmatic radio sources known as quasars, by means of their scintillation or twinkling.

Every amateur astronomer knows that you can easily tell a planet from a star because a star, being a point of light, usually twinkles whereas a planet doesn't. Twinkling is caused by unsteadiness in the Earth's atmosphere which switches the thin beam of light from a point-like star back and forth over the pupil of the eye. A planet twinkles much less, if at all, because it has a disc, even though it may seem the same size as a star.

Much the same sort of thing happens with radio sources, though radio astronomers use the word *scintillation* instead of twinkling. The medium causing the disturbance is thin ionized gas called plasma. Not only is plasma found in the atmosphere of the Earth but also far out in the solar system, between the planets.

For Antony Hewish and his team, who first picked up interplanetary scintillation from point-like radio sources in 1964, its importance was that it enabled the tiny, distant quasars (quasi-stellar radio sources) to be picked out from nearer, apparently larger, sources.

To get the sensitivity necessary to distinguish rapid fluctuations of signal, Hewish needed a radio telescope with a large collecting area. This he achieved by simply setting up wires on poles, covering a paddock of 4½ acres. This had the required ability to pull in faint signals, but it lacked the direction-finding discrimination of a dish-type antenna. By July 1967, the new telescope was ready to begin recording. It was designed to scan a large part of the sky in 1 week. The equipment was built to emphasize the scintillation rather than, as was usual, to de-emphasize it; it was able to respond in as short a time as one tenth of a second to fluctuations in the brightness of a radio source. Hewish wanted all the radio sources which it found to be plotted on a map of the sky, so that terrestrial man-made interference, which would appear randomly, could be sorted out from the truly extraterrestrial twinkling radio sources which would recur at the same celestial coordinates. The person to whom he assigned the job of analysing the data from the instrument was a graduate student, Jocelyn Bell.

Scruffy little green men?

Analysing the data from the new telescope was no small task. The instrument produced 400 feet of tape from each scan of the sky, 100 feet every day. Bell's job was to examine every signal, discarding such man-made phenomena as aircraft transmissions and foreign television stations and mapping out the true extraterrestrial signals. By October, she was 1000 feet behind current chart production and yet, fortunately, she did not relax her standard of attentiveness.

It was in October that Bell noticed what she called 'a bit of scruff'. It was passing through the beam near midnight, when the interplanetary scintillation normally falls to a low level (because, at this time, the radio telescope is pointing to the outer edge of the solar system where the plasma is least dense). Bell's account says:

Sometimes within the record there were signals that I could not quite classify. They weren't either twinkling or manmade interference. I began to remember that I had seen this particular bit of scruff before, and from the same part of the sky.

The source seemed to be recurring every 23 hours 56 minutes, and only objects fixed among the stars recur every 23 hours 56 minutes. Man-made interference, though, tends to recur on a 24 hour schedule, because daily life is ordered by the Sun. The moment when Bell recognized that the bit of 'scruff' was more than a single piece of interference, but had actually occurred before at the same celestial coordinates, proved to be very important. Bell describes her reaction:

When it clicked that I had seen it before I did a double take. I remembered that I had seen it from the same part of the sky before.

Looking back at the records, Bell was able to prove that she had, in fact, seen it 2 months before, in August. She then discussed the signals with Hewish. They decided to use the observatory's fast recorder to get a clearer picture of the nature of the signals. When the fast recorder became available in mid-November, Bell was given the job of trying to catch the signals and record them. For some days she was unsuccessful. At this point, Hewish thought that the signals were from a randomly occurring flare star and that it was unlikely that Bell would see it again. She persevered, however, and at last managed to catch a satisfactory recording which showed clearly that the 'scruff' was a burst of pulses almost exactly $1\frac{1}{3}$ seconds apart, similar to many kinds of terrestrial interference. When she telephoned to Hewish to tell him what she had found, he said 'Oh, that settles it. It must be man-made.'

Nonetheless, Bell and Hewish continued to make recordings of the 'scruff'. The main problem was still that the pulses were keeping sidereal time (recurring with the 23 hour 56 minute period). Were the bursts truly sidereal, or were they being made artificially with a sidereal period? The only people on Earth who could conceivably be imitating sidereal time would be astronomers, though no one could guess why they would want to make bursts of pulses like this. Inquiries at other observatories failed to reveal any programme which could account for the signals. Searching around for sidereal explanations for the pulsing signal, Hewish and Bell considered whether known variable stars could cause it. The trouble was that the fastest variable star known had a period of about one-third of a day. How could a star throb with a period of 1.337 seconds?

Caught in the dilemma that the pulses were extraterrestrial, but seemed to be artificial, the

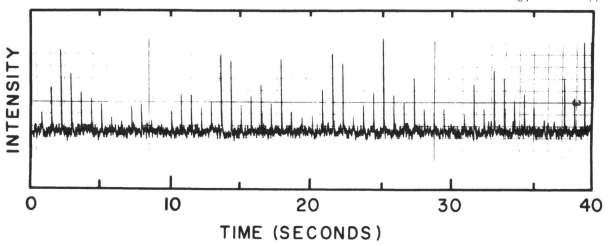

FIG. 40. *Radio pulsar. This is a recording of the signals from radio pulsar PSR 0329+54, observed by Richard Manchester with the National Radio Observatory's 92-metre transit telescope. Each spike on the chart is a pulsar radio burst. In this pulsar the bursts recur at 0.714 second intervals.*

Cambridge astronomers began to consider a new possibility: were the pulses being manufactured in space by an extraterrestrial civilization? By mid-December, they had proved that the pulses recurred very regularly indeed, staying on schedule to one millionth of a second. In a half-joking way, Bell's colleagues began to refer to the star as LGM-1. But why would Little Green Men manufacture and broadcast repetitive signals like this? Most radio signals change in order to convey information; the constant ones are navigational aids like the LORAN signals (for LOng RAnge Navigation). Was LGM-1 an interstellar navigational beacon?

If the pulses were being manufactured artificially by an intelligent civilization, the manufacturers presumably lived on a planet. If the signals were coming from a planet, they would show the effects of a *Doppler shift*. The Doppler shift causes a bunching effect of repetitive signals as the transmitting object moves towards the recipient, and a spacing out effect as it moves away. The Cambridge astronomers had, in fact, already observed a small change in the timing of the signals, caused by the motion of the Earth around our Sun. Could they see the equivalent effect, at a different period, caused by the transmitter itself being on another planet orbiting its own sun? Bell recorded in her log,

We are working on the Doppler shift of the pulses to see whether the source is stationary or moving round a sun. There is no 4C [Cambridge catalogue] source with the same coordinates, nor any other source that we know of.

In the event, no Doppler shift was seen, other than that caused by Earth's motion. The radio source was therefore not on another planet. The little green men became less likely.

The Cambridge radio astronomers used the amount of Doppler shift on the period of the pulses to estimate the source's position in the sky because, as we have noted, the 4½ Acre Telescope lacked direction-finding precision. Suppose that a pulsar with a period of one second lies in the plane of the Earth's orbit. On a certain day of the year, the Earth will be travelling directly towards it with its orbital speed of 30 km/s. This will cause a decrease of the period by one ten thousandth of a second to 0.9999 seconds. 6 months later, the Earth on the other side of its orbit will be travelling away from the pulsar at 30 km/s, and the period will be lengthened to 1.0001 seconds.

On the other hand, if this imaginary pulsar is in a direction perpendicular to the Earth's orbit (a direction known as towards the pole of the ecliptic), the Earth will throughout the year be travelling across the line of sight to the pulsar, no Doppler shift will be observed and the period will always be 1.0000 seconds.

Such small changes may seem impossible to measure, particularly when we see the indistinct signal from a weak pulsar. But we are able to take the average of hundreds or thousands of pulses. After 10 000 pulses of the hypothetical pulsar, each one delayed by 0.0001 seconds, the combined delay will be 1 whole second. 10 000 seconds is nearly 3 hours, so in a brief space of time it is possible to measure the rate of a pulsar to much better than 1 part in 10 000. From the size of the Doppler shift of the first pulsar, the Cambridge team estimated its position and that it was unmoving, to an accuracy of about 2 arc minutes. Just as Brahe had shown for the supernova of 1572, and Kepler for the supernova of 1604 (and with the same accuracy), the Cambridge astronomers had demonstrated that the pulsar had no discernible parallax and must be farther than the edge of the solar system.

Measuring the pulsar's distance

The Cambridge team then proceeded to make some measurements which were possible with two radio receivers operating simultaneously, but tuned to different radio frequencies.* Pulses observed on the two frequencies arrived at different times, a pulse travelling on the longer wavelength radio arriving later than the same pulse travelling on the shorter wavelengths. The delay between the two signals showed that the radio frequencies travelled through interstellar

space at different speeds, a phenomenon called *dispersion*.

If interstellar space were truly empty, all radio waves would travel at the speed of light, but there are free electrons in interstellar space, produced from the ionization by starlight of atoms of interstellar gas, such as sodium. Starlight passing near a sodium atom can interact with the atom, causing it to eject the loosest of its electrons. Interstellar space is thus not empty: it contains a plasma, a low-density gas containing unattached electrons, so radio waves travelling through it (especially low wavelength ones) are slowed from the speed of light by tiny amounts.

The amount by which the radio waves are slowed down depends on the density of electrons in the plasma. In interstellar space, radio waves are slowed down by typically 1 inch/second from the 186 000 miles/second speed of light. The time delay which this causes between pulses observed at different radio wavelengths is called the *dispersion measure*, and depends on the square of the electron density multiplied by the distance to the star. By assuming a value for the electron density in space (it had previously been studied by other means), the astronomers were able to estimate the distance of the pulsating radio source to be 200 light years, placing it among the stars rather than close to the solar system or outside the Galaxy altogether. Though the slowdown of radio waves by the interstellar plasma is a small decrease from the speed of light, the distance that the waves travel is so large that the difference in speed causes a measurable delay of, typically, 1 second.

How big is it?

At this time the radio astronomers were also able to make an estimate of the size of the pulsating star (it was this latter phrase which was

---

* The frequency and wavelength of radio signals – or any other kind of waves – are linked. The frequency is the number of waves passing a point in a given time, while the wavelength is the crest-to-crest distance of the waves. As the wavelength gets shorter, so the waves get more frequent and the frequency is higher.

contracted to *pulsar*). They measured the length of the individual pulses at a particular radio wavelength and found that each pulse lasted for about 16 milliseconds, only 2% of the period between the pulses. Each cycle of the pulsar was a brief flash with a relatively long time between flashes; whatever was causing the flash had to emit the light all within a time of 16 thousandths of a second.

The duration of the flash arose from two causes. Consider two parts of the flashing region: let us suppose that each part simultaneously makes a brief mini-flash, but let us suppose that the two parts are separated on our line of sight by a distance, say *d*. At the speed of light, *c*, the rear flash takes an extra time *d/c* to reach Earth. Therefore when the flashes are observed at Earth they appear as a pair of pulses separated by an interval equal to the time taken by light to travel the distance between the two parts of the flashing region.

Clearly, if the flashing region has many components spread over a distance along the line of sight, the flash observed would be smeared out over the time that would be taken by light to travel that distance. If the flashes from the component parts are not really simultaneous then the spread of the pulse will be even more than the light-time along the depth of the source. Therefore, the duration of the flash of the pulsar as observed on Earth shows the maximum extent of the pulsar's depth. The pulsar observed by Bell and Hewish must be smaller than 16 light milliseconds in depth: less than 3500 miles.

What sort of object could emit rapid, energetic radio pulses, yet be smaller than the Earth? This is much smaller than any ordinary star, but it is about the size of certain very condensed stars. For many years, astronomers had been aware of the existence of compact stars.

Among them are the white dwarfs (the end products of stars in which the normal energy processes have ended, leaving them at the mercy of their own gravity). These stars collapse upon themselves, forcing their atoms into a super-dense state called *degeneracy*, and resulting in the entire mass of the star being packed into a body little larger than the Earth. White dwarfs are not uncommon in the sky – one is a companion of the bright star Sirius, for example. But theoretical astronomers did not rest content with white dwarfs. There should be even more compact stars, they said, smaller and denser than white dwarfs. They called them neutron stars. For 30 years, neutron stars were the theoretical solution to a problem which did not exist: mathematically it had been shown that they could exist but no trace had been found of a real example. As evidence grew that the newly discovered pulsar was a natural phenomenon, the thoughts of the Cambridge astronomers turned to stars of this kind. The rotation of a neutron star was capable of providing the repetitive clock which could pace the pulses. There were, however, still more discoveries to be made.

More pulsars turn up

In December, Bell discovered a second pulsar.

I was working in the evening analyzing charts. I saw something which looked remarkably like the bits of scruff we had been working with. This was in a bit of sky which wasn't very easy for the telescope to look at, but there was enough to confirm that there had been scruff.

That particular bit of sky was due to go through the beam at one in the morning. It was a very cold night and the telescope doesn't perform very well in cold weather. I breathed hot air on it, I kicked and swore at it, and I got it to work for just five minutes. It was the right five minutes, and at the right setting. The source gave a train of pulses but with a different period of about one and a quarter seconds.

Finding a second pulsar made it even less likely that the transmissions were artificially produced by another civilization.

There wouldn't be two lots signalling us at different frequencies. So obviously we were dealing with some sort of very rapid star. I threw up another two sometime in January.

'Throwing up' another two had involved searching back through the 3 miles of tape which had accumulated. Now that they had this further evidence that the signals were from a natural galactic source, the astronomers felt ready to publish the news of the discovery of the first pulsar. They wrote the paper describing the observations, and submitted it to the science journal *Nature*, 8 weeks after the recognition that the first radio source was pulsating. Shortly afterwards, on 1968 February 24 the paper on the first pulsar was published in *Nature*, by Hewish, Bell, and their colleagues J. Pilkington, P. Scott and P. Collins.

The story of the discovery of pulsars has a pattern which recurs regularly in the history of science, though not as repetitively as pulsars themselves. Built for a completely different purpose, Hewish's radio telescope picked out the pulsars by chance. Bell noticed the signal from the first pulsar because, though it resembled both spurious interference and the normal twinkling of radio stars, its characteristics did not quite fit with either because it recurred in the same part of the sky and it twinkled at midnight. She was still a graduate student and therefore inexperienced in astronomy, but she had a receptive mind and seems to have been less ready to dismiss the

'scruff' as interference than her more experienced colleagues (there is a persistent rumour that pulsars had been seen by a radio astronomer before Bell, but had been dismissed by him). After patiently but quickly bringing out the essential characteristics of the pulsars, the Cambridge group, armed with the confidence that they had discovered further examples, published their data and were able to mention in their discussion of it, what is now believed to be the kind of star responsible.*

A row flares

Like many stories of superb scientific discoveries, this one has parts that could have been scripted by C. P. Snow. The Cambridge group was criticized for sitting on the discovery of the first pulsar for 6 months and then concealing the discovery of the further pulsars which they had found. Actually there were only 2 months between the recognition of the repetitive pulses from the first pulsar and submission of their paper to the journal *Nature*.

Professor Sir Bernard Lovell, Director of Britain's Jodrell Bank Observatory said he thought

---

* Fred Hoyle's novel *A for Andromeda*, written with John Elliot and published 5 years before Bell's discovery, contains a remarkable foreshadowing of the discovery of pulsars. A large, new radio telescope picks up from the direction of the constellation Andromeda a 'faint single note, broken but always continuing' like Morse code. The radio astronomers deduce from its constant position in galactic coordinates that it is not in orbit in our solar system, nor an artificial satellite. Unlike the real pulsars the signal turns out to be a message, with a lot of 'fast detailed stuff' between the dots and dashes. Like the Cambridge radio astronomers, the fictional ones build a fast recorder to record it. A news blackout is imposed by a high-up civil servant, to be broken by the most individualistic of the scientists. The press sensationalize the signal: SPACEMEN SCARE: IS THIS AN ATTACK? Perhaps this is another reason why, when pulsars were first discovered at Cambridge, the radio astronomers imposed a news blackout on themselves, avoiding misrepresentation when they were unsure of precisely what they were observing.

the Cambridge people had behaved with exemplary scientific discipline in withholding news of their discovery until they were satisfied about the general nature of the objects, but that there was no excuse for a similar delay in withholding information about similar objects which they had discovered.

But this incident was of small importance compared with a later row.

In 1974, the Nobel Prize for physics was awarded for the first time to astronomers. It was given jointly to Martin Ryle, Director of the radio observatory at Cambridge for his work in developing new kinds of radio telescopes, and to Antony Hewish for his 'decisive role in the discovery of pulsars'. The Nobel Prize committees are thought to investigate the circumstances very thoroughly when they make their awards, and they seem to have been satisfied that Bell's part in the discovery did not merit a share in the prize.

In the following March, however, the well-known British astronomer, Fred Hoyle, one-time director of the Institute of Theoretical Astronomy in Cambridge, criticized the way in which the Nobel Prize had been awarded to Hewish without including Bell. According to Hoyle, the crucial parts of the discovery were the recognition of the signal as something unusual, and the observation that it was keeping sidereal time. After this, any astronomer would have gone through the same reasoning process and come to the same conclusions as the Cambridge team. Hoyle wrote:

There has been a tendency to misunderstand the magnitude of Miss Bell's achievement, because it sounds so simple, just to search and search through a great mass of records. The achievement came from a willingness to contemplate as a serious possibility a phenomenon that all past experience suggested was impossible. I have to go back in my mind to the discovery of radioactivity by Henri Becquerel for a comparable example of a scientific bolt from the blue.

In a reply Hewish wrote that Bell had been carrying out a programme initiated and mapped out by him. He said that her work as a graduate student had been excellent, but that it would be unjust to later graduate students who continued the analysis to suggest that they would not have discovered the pulsar themselves, had they been in her position. Of course, the argument that the next person down the line would have made the discovery anyway if X had not, applies to most scientific work. Very often in great discoveries there is an element of luck: being in the right place at the right time, with the right predecessors. No matter who received the Nobel Prize, Bell actually discovered the first pulsar of the more than 300 now known.

More discoveries

After the four Cambridge pulsars were found, others were discovered by astronomers using radio telescopes at Green Bank, West Virginia; Jodrell Bank; Arecibo, Puerto Rico; and Molonglo in Australia. In a discovery late in 1968, which indicated the connection between pulsars and supernovae, a Sydney University group using the Molonglo radio telescopes discovered a very short period pulsar which lay in the same direction as a source of radio emission called Vela X. Could the two be linked? Vela X had been previously identified by Douglas Milne as the remnant of a supernova which occurred some 10 000 years ago. Milne put Vela X at a distance of about 1700 light years, and the Sydney radio astronomers M. I. Large, A. E. Vaughan and Bernard Mills deduced from its dispersion measure (p. 78) that the pulsar which they discovered was at the same distance. They inferred that the pulsar probably was the stellar remnant of the supernova – the small star left over after the supernova explosion which had ejected the outer parts of the original star into space and made the Vela X radio source. The pulsar had a very short period – only 89 thousandths of a second, and each brief flash of radio waves lasted only 10 thousandths of a second.

Within a few weeks, however, an even shorter-period pulsar (still the shortest period known) was found in the centre of the Crab Nebula which proved the connection between pulsars and supernovae. This one was discovered in a deliberate search of the Crab for its pulsar by D. H. Staelin and E. C. Reifenstein at the National Radio Astronomy Observatory in Green Bank, West Virginia, using a brilliant method which exploited one of the properties that made its detection as a pulsating star difficult – its high dispersion.

As explained previously, pulsar pulses, when observed at different radio wavelengths, arrive at Earth at different times. No radio telescope observes at a single radio wavelength – it always detects radio waves arriving within a small band of wavelengths, called the bandwidth of the radio receiver. (This is true of a domestic radio receiver too: cheaper quality radios receive a wider bandwidth than more finely tuned, expensive ones and may receive the programmes broadcast by two radio stations at adjacent wavelengths on the wavelength dial. The two stations may interfere with each other and their individual signals are muddled.)

The wider the bandwidth in a radio astronomy receiver, the more radio energy it receives from the sky and, as a consequence, it can detect fainter radio stars. If a radio star is pulsating fast and there is strong dispersion, a pulse from the pulsar can be received at the lower wavelength edge of the bandwidth at the same time as previous pulses are being received at longer wavelengths still in the receiver's bandwidth. As a result the pulses are blurred together and not distinguished or recognized as pulses by the radio telescope.

Staelin and Reifenstein realized that this would occur if they were to search for a pulsar in the Crab Nebula since dispersion is caused by electrons in space and the Crab contains many free electrons. This is evident from its appearance, with red H-alpha filaments and overall synchrotron glow. They decided therefore to use finely tuned narrow bandwidth radio receivers so that they could observe individual pulses, compensating for each receiver's individual lack of sensitivity by coupling them all together in a bank of 50, and looking for a particular pulse as it was received in turn by each of the receivers. It turned out, when they in fact discovered the Crab Nebula pulsar, that it took 1½ seconds for a pulse from the pulsar to sweep through all 50 of their receivers, because of the dispersion. The period of the pulsar turned out to be just 33 thousandths of a second.

The two shortest period pulsars then known had been discovered to be associated with supernova remants. Searching for a reason for this, astronomers hypothesized that the youngest pulsars had the shortest periods and that pulsars slow down as they age. Only young and, hence, short-period pulsars would be found associated with supernova remnants, as the remnants associated with old pulsars should have dissipated into space, fading from view.

Almost instant confirmation of this idea came from the discovery that the Crab pulsar was not completely regular. Within 1 month the period had been measured accurately enough to demonstrate that it had increased by 1 millionth of a second. This steady increase meant that the Crab Nebula pulsar was slowing down at a rate which indicated a lifetime of just about 1000 years – near enough the age of the Crab Nebula

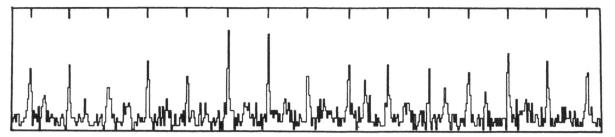

FIG. 41. *Optical pulsar. The graph shows observations of the light flashes from the Crab pulsar made with the Anglo-Australian Telescope. The main flashes, marked on the top scale, are 0.033 seconds apart. Between each pair of flashes can be seen the so-called interpulse, smaller flashes from a weaker beam on the rotating neutron star, following its main beam by 0.013 seconds. Doubled flashes like this are relatively common on the faster pulsars.*

as determined from the Chinese observations of the supernova of 1054! By the time this was known late in 1968, the four pulsars discovered by Bell and her Cambridge colleagues had been under observation both at Cambridge and Jodrell Bank for more than a year, and their periods were becoming known with greater and greater precision. All four were found to be slowing down, but some 10 000 times more gradually than the Crab pulsar. Not a single one of the 330 radio pulsars known is speeding up – they are all either constant or slowing down – further indication that the rotation of pulsars slows as they age.

Seeing the Crab pulsar flash

Although radio observations immediately showed clearly that the pulsar discovered by D. H. Staelin and E. C. Reifenstein was in the Crab Nebula, its position could not be measured well enough with radio telescopes to determine precisely which star was the pulsar. Indeed, although many attempts had been made to identify the first radio pulsars with particular stars – and there were several false alarms – no radio pulsar had been found to be identical to any optically visible star. These earlier disappointments may explain the casualness of the efforts made by optical astronomers with access to large telescopes to identify the Crab Nebula

pulsar. One astronomer used the 98-inch Isaac Newton Telescope in Britain, 6 days after the periodicity of the radio Crab pulsar was announced, to observe light from the central regions of the Crab Nebula, but left his data (which was in the form of punched tape readable only in a large computer), unanalysed for months with no inkling of the discovery hidden within.

Then, a team of three astronomers and physicists at the Steward Observatory in Tucson, Arizona used the much smaller 36-inch telescope there in January 1969 to look in the centre of the Crab Nebula for light flashes with the radio period. In their experiment, W. J. Cocke, M. J. Disney and D. J. Taylor used a small computer to add thousands of the flashes together, and – this turned out to be a significant difference – they had the results of the computer summation available to them in graphical form then and there, while the experiment was in progress. As the pulses came in from the Crab pulsar, they were displayed on a multi-channel recorder – a glowing screen on which pulses appeared as a hump on a line, growing as the astronomers watched. They were able to sweep the telescope over the central regions of the Crab Nebula to see where the pulses came from and were able to say that they came from the vicinity of the two central stars, the northeast and southwest stars named by Baade. They could not say which, for sure, but guessed the southwest star as this had been the one backed by Baade and Minkowski.

The Steward Observatory experimenters' guess was confirmed in a clever experiment by two astronomers from the Lick Observatory, Joe Miller and Joe Wampler. They used a TV camera attached to the 120-inch Lick telescope, peering at the pulsar through a rotating shutter which was made so that the open periods were separated by a time very nearly the same as the period of the pulsar. Similar devices, called stroboscopes, are used to view rapidly rotating machinery; the stroboscopic effect is also familiar as a slowing of the motion of a wagon wheel when filmed by a movie camera which exposes film frame by frame, through a shutter. When the pulsar was flashing at the same times that the shutter was closed, the Lick astronomers saw nothing. But when it flashed at the same time as the shutter was open they immediately saw that the pulsar was Baade's southwest star, right at the centre of expansion of the Crab, and at the centre of activity of the light ripples seen by him.

Astronomers had to look no further for the powerhouse which drives the Crab's expansion and generates the electrons to produce its synchrotron radiation.

An exciting moment

If a pulsar did not exist in the Crab Nebula it would be necessary to invent one. Indeed it was necessary, for the year before the discovery of pulsars, F. Pacini suggested that a rotating neutron star was the present-day source of the fast electrons in the Crab Nebula. The discovery of the Crab Nebula pulsar showed that this was possible, and the further discovery that the pulsar was slowing down proved it, as the star's spin-

energy lost in the slowdown was closely equal to the total amount of light and radio energy radiated by the synchrotron process. The whole flow of energy within the nebula became clearer.

Thomas Gold of Cornell University recalls the moment when he too realized that the Crab Nebula was powered by a rotating neutron star:

Let me just recount for you our excitement at Cornell when we had just obtained, at our observatory at Arecibo, a slowdown rate for the Crab for which we had been looking. We had expected a supernova to give rise to a neutron star, which if found, was expected to be slowing down. Therefore, when the pulsar in the Crab was discovered, we immediately started looking for a slowing down of the pulsations since such a short period pulsar should slow down fast. When we got the information, we immediately worked out the rate of change of energy of a rotating neutron star, having been previously very impressed with the very high total energy content in the rotation of the object.

I remember doing the completely simple calculation of what the rate of change meant in terms of the energy output, and meanwhile sending an assistant to the library because I no longer remembered the quoted figures for the total energy output of the Crab Nebula that had been calculated by Shklovsky years before. When he returned in a few minutes with various references giving estimates of the luminosity of the Crab we found that the figure on my pad and the other in the book were the same, namely 10^{35} ergs per second. That was a really exciting moment. I realize that we can't be quite sure that this is the right number, but still, it isn't often that you hit it off like that with a completely theoretical calculation of such a far-fetched thing as the structure of a neutron star.

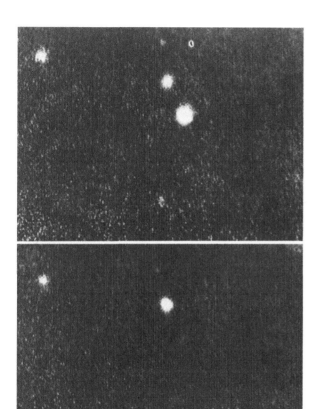

FIG. 42. *Television pulsar. This pair of images of the Crab Nebula pulsar was taken with the Lick Observatory 120-inch telescope using a television system looking through a stroboscopic shutter. The shutter was set to rotate almost in synchronism with the 0.033 second period of the pulsar. The top photo was exposed off a TV monitor when the pulsar was at its brightest, during its flash. The lower was made when the pulsar was almost invisible. The other stars nearby have a constant brightness. It was this pair of photographs which conclusively demonstrated that the pulsar was the southwest of the pair of stars at the centre of the Crab Nebula. This is a Lick Observatory photograph by E. J. Wampler and J. Miller.*

Superstar . . .

There was however still an unexplained problem. The pulsar itself was too faint. Responsible for the vast amount of energy pouring into the Crab and manifest as accelerating filaments and as synchrotron radiation, the pulsar itself emitted just 18 millionths as much energy as visible radiation. It was scarcely credible that the pulsar would pulse on its own behalf such a small amount of energy, while passing on such huge amounts, any more than it is credible that a showbusiness superstar would accept a small paypacket while generating millions of dollars worth of business. Searching for the Crab pulsar's piece of the action, NRL and Columbia University X-ray astronomers launched two rockets in March 1969 and found X-ray pulsations. This prompted other astronomers to search through and re-analyse old data which they had not previously thought to examine for X-ray pulsations. These pulsations occur exactly in step with the optical flashes, and the period of both is the same.

The energy which the Crab pulsar puts into X-radiation is 20 times the energy it emits optically and 20 000 times the energy it pulses as radio radiation. This energy is about 5% of the total energy of the Crab Nebula; a significant fraction of the spin-down energy of the pulsar therefore goes into pulsating X-rays.

One interesting question is how bright the Crab pulsar is when it is 'off'. The pulsing X-rays represent bright beams from the pulsar and, when the beams are pointed away from the Earth, it may be possible to glimpse the fainter hot surface of the neutron star. In fact, nothing from the hot surface can be seen. Observations with the Einstein Observatory Satellite show that the maximum temperature of the Crab pulsar's surface is 2.5 million degrees K. The observation is potentially useful to nuclear physicists who can attempt to calculate the thermal conductivity of neutron star material. The pulsar must have been born hot and cooled to this temperature, or below, during its 1000 year lifetime. Only if the conductivity of neutron star material is good enough can the heat in the star flow out fast enough. Since neutron star material is completely unknown on Earth the conductivity can be estimated theoretically only. The calculations are esoteric, and depend on the structure of the neutron star, the proportion of neutrons to other nuclear particles (like pions or quarks), the properties of the crust of the neutron star and the strength of its magnetic field. There are some problems about the assumptions under which the calculations are applied – for instance, one has to make the assumption that nothing has heated the surface of the neutron star during the pulsar's lifetime – but so far the nuclear physicists are happy with themselves at getting answers which are consistent with the observation. This encourages them to believe that their understanding of the nuclear physics of neutron star material at the extreme pressures and densities in neutron stars is not completely off target.

... and co-star

Until 1975, the Crab pulsar was the only pulsar detected as an optical or X-ray pulsar as well as a radio pulsar. The search for other examples had proved fruitless. The most likely pulsar to be discovered as an optical or X-ray pulsar was the Vela pulsar in the Vela supernova remnant. Before 1975 it was the second fastest pulsar known and so had more rotational energy to convert to light and X-rays than any other pulsar apart from the Crab. X-rays have in fact been detected from the Vela pulsar, and there was one report in 1973 that the X-rays pulsed, but more sensitive X-ray

telescopes failed to see any pulsations – just a small cloud surrounding the pulsar position. In 1975 however, the Small Astronomy Satellite, SAS-2, detected pulses from the Vela pulsar at very energetic X-ray energies (gamma rays, in fact).

Optical astronomers redoubled their efforts to find visible light from the Vela pulsar. Though 4 times nearer than the Crab and in a less dusty region of the Galaxy, the Vela pulsar was expected to be much fainter than the Crab because it was 10 times older and 3 times slower. The optical search was hampered by the disagreement among radio astronomers about the precise direction of the pulsar. Within the zone of uncertainty there were too many faint stars to be examined in the detail necessary to pick it up. However after radio astronomers in Australia had pinpointed the position of the Vela radio pulsar with an accurate radio telescope at Fleurs, near Sydney, they joined forces with several optical astronomers from Australia and the UK. In 1977, using the Anglo-Australian Telescope, this large team found the Vela pulsar after 10 hours integration on the right place in the sky. Light flashes from the Vela radio position were stored in a computer, integrated spontaneously in an analysis at the telescope, and remembered for subsequent more refined analysis afterwards. The Vela pulsar, 10 billion times fainter than bright naked-eye stars, is one of the faintest stars to be seen by optical astronomers.

In later observations, the same team was able to obtain pictures showing the pulsar flashing in its cycle. They used a television detector called the Image Photon Counting System. The image of the pulsar in the Anglo-Australian Telescope was intensified electronically and detected by a television tube, which created a small television frame every 6 milliseconds. The frame was dealt to one of eight stacks in a computer memory, with the computer routing the frame to the stacks in a cycle which was in sympathy with the pulsar period. As each stack accumulated frames, it developed a picture which was essentially an image of the pulsar frozen at one of eight moments throughout its cycle. The method was a digitally controlled version of the stroboscopic method used to see the Crab Pulsar at the Lick Observatory in 1969.

Although no X-ray pulsations have been seen from the Vela pulsar, the pulsar position is surrounded by an X-ray nebula, about 1 degree in extent. It seems likely that this nebula represents synchrotron emission from electrons ejected from the pulsar, much like the white continuum radiation from the Crab Nebula.

Two more pulsars are known in supernova remnants. An X-ray/optical pulsar with 0.05/s period lies in a Crab-like supernova remnant in the Large Magellanic Cloud, and there is an X-ray and radio pulsar lying within a remnant called MSH 15–52. Both pulsars power synchrotron nebulae which surround them. These pulsars which are associated with supernova remnants raise a general question about the connection between pulsars and supernovae.

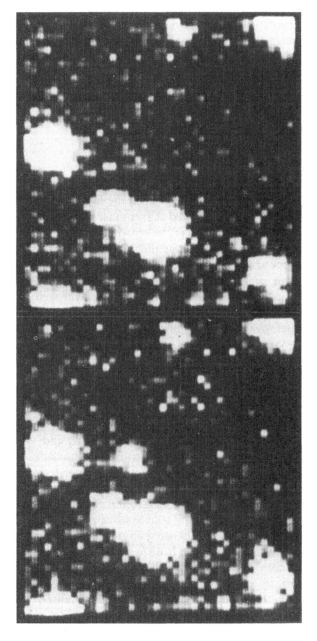

FIG. 43. *Computer pulsar. Like the previous figure, this pair of pictures shows an optical pulsar 'off' (top) and 'on' (bottom). Because the Vela pulsar is 10 000 times fainter than the Crab pulsar, the images were added in a computer system rather than made visible by direct TV techniques. The individual pixels (points in the picture which are recorded in the computer memory) show as small squares, and noise associated with the faint image shows as a sprinkling of 'snow' in the dark areas. But the pulsar is visible in the bottom picture as a collection of pixels much brighter in the bottom picture than in the top: it lies near the centre of the frame, above the merged images of two stars. This photograph from the Anglo-Australian Telescope.*

7

Supernova remnants

When a star goes supernova, it leaves traces. It is not literally true that these remains are the dead body of the star because, in the case of the Crab at least, the remains are still active, emitting as much energy as a luminous star. Are there other supernova remnants? What about Tycho's and Kepler's stars, and all the other historical supernovae?

Renaissance supernova remnants

After Walter Baade had studied the light curve of Kepler's star, he realized that it must have been a supernova and, in 1947, he searched for its remnant. Using what was then the world's largest telescope, the 100-inch at Mount Wilson, he photographed the suspect area using a red filter to isolate the red light expected from the hydrogen in the nebula. At the position of Kepler's supernova he found a few wispy filaments of gas, the last traces of the exploding star of 1604.

Baade applied the same principles to a search for the remnant of Tycho's supernova of 1572. He was unsuccessful – photographs centred on Tycho's position for the supernova showed stars but not a nebula.

Then in 1952 Robert Hanbury Brown and Cyril Hazard searched for Tycho's supernova remnant with a 218-foot radio telescope at Jodrell Bank (not the fully steerable 250-foot telescope but a simpler telescope fixed to the ground). They found a powerful radio source which, according to a Cambridge group using an interferometer, was in the right place. It turned out that the position Tycho gave had the unexpectedly large error of 4 arc minutes – over a tenth of the diameter of the Moon. Though this had caused Baade to look somewhat off the correct place for the nebula, the real position was still on the edge

of his plates and he would probably have seen the nebula if it were bright enough.

However, the remnant of Tycho's supernova is very faint indeed and photographs by R. Minkowski with a 200-inch Mount Palomar telescope show just a few wisps of nebula near the radio source. The radio source has been mapped by the 1-mile Cambridge radio telescope. It turns out to be beautifully circular in shape with a ring of bright radio emission, the edge of a bubble of material thrown violently into the interstellar gas surrounding the supernova, colliding with it, and heating the gas to a billion degrees. This is why so little light is seen – the gas is too hot for hydrogen atoms to form and, in doing so, emit light.

No astronomer doubts that this is the remnant of Tycho's supernova. Why then is it so far from the position Tycho measured for the star? The measurements that he made on the nova with his new sextant extended over many months, but there is no record of Tycho using it for any of his subsequent observations, known to be very accurate. According to astronomer David H. Clark and historian F. Richard Stephenson the discrepancy can be explained purely as a small calibration error in the angular scale of the sextant – as if Tycho were measuring metres with a yardstick.

Once the remains of the supernovae of 1054 (the Crab Nebula), 1572 (Tycho's supernova) and 1604 (Kepler's supernova), had been identified with the wispy supernova remnants, astronomers began to look for other examples. They were hoping to find the remnants of supernovae which had not been recorded, either because they were too faint to be seen or because they occurred in prehistoric times so that no records of the supernovae have come down to us.

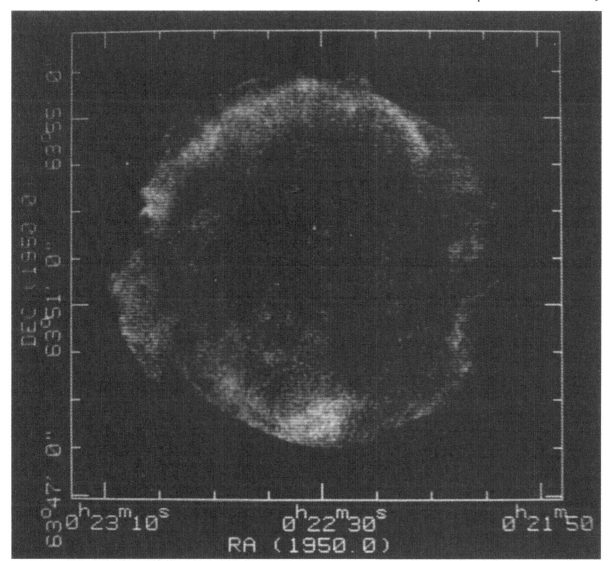

FIG. 44. *Tycho's supernova remnant. Radio telescope pictures of supernova remnants can be sharpened in a computer by a Maximum Entropy Method to recover the finest detail. The same technique (first developed by radio astronomers) can be used to sharpen indistinct TV pictures so as, for example, to read the blurred licence plate number of a speeding car. This radiograph of Tycho's supernova remnant, 3C10, is by Dave Green and from the Mullard Radio Astronomy Observatory, Cambridge. It shows a thin shell of radio emission departing only slightly from sphericity. The small bumps and dips on the edge of the shell represent places where the expansion is encountering a somewhat lumpy interstellar medium, the denser parts of which slow down the outward motion of the shell more than the more rarefied parts. The faint optical filaments of Tycho's remnant are hard to see through the absorption of interstellar dust along the line of sight to the supernova (see Fig. 57), but lie near the brightest radio regions at the north and south of the shell (top and bottom) and in the indentation to the east (left).*

The radio traces of ancient supernovae

In the cases of the well-established historical supernovae, radio sources were observed at their positions. The characteristics of these radio sources were used to define the appearance of a radio supernova remnant, and then radio astronomers looked for other similar sources in order to make a complete supernova-remnant inventory. Radio waves penetrate through the interstellar dust, so that radio astronomers can survey the whole of our Galaxy, even where optical astronomers are frustrated by interstellar obscuration, just as an air-traffic controller can use radar to penetrate fog when aircraft cannot be seen. Radio astronomers have detected many more supernova remnants than optical astronomers have. The latest catalogue of supernova remnants lists 135. Forty of these have been detected by optical astronomers and 33 by X-ray astronomers.

To a radio astronomer, a supernova remnant is an 'extended, non-thermal galactic source'. What do these adjectives mean? Why does this definition connect with supernovae?

Most radio sources are extragalactic – they are quasars and galaxies far outside our own Galaxy. Quasars themselves are star-like – when studied at high resolution they may be double, or show jets or haloes, but basically they appear as points on the radio sky. Galaxies are extended – or they have a shape – but because most are distant they appear small and, again, basically point-like. With the exception of a few very nearby galaxies (like the Andromeda galaxy), the only objects which show appreciable extent in a typical small radio telescope are inside our Galaxy. These then are the 'extended, galactic radio sources'. What about the adjective 'non-thermal'?

There are three fundamental sorts of extended, galactic radio source: clouds of cold interstellar hydrogen, clouds of interstellar hydrogen heated by stars and supernova remnants. They are distinguished by their radio spectra. The cold interstellar hydrogen emits a powerful radio signal at a wavelength of 21 cm, and is unmistakable. Nebulae and supernova remnants have spectra which are much less specific. Radio astronomers measure the intensities of the radio sources at several radio wavelengths and plot their results on a graph. They look at the shape of the graph produced. This shape indicates the way that the radio radiation is produced. Nebulae heated by stars contain ionized hydrogen – free protons and free electrons. The electrons move with speeds of a few thousand kilometres per second, which is fast by human standards but slow for an electron, which could reach nearly the speed of light (300 000 km/s). The so-called thermal electrons wander over the nebula, encountering and mutually repelling one another, changing their motion at random. Whenever an electron changes its motion it emits radiation, so that the nebula gives off radio waves. The motions of the electrons and their encounters depend on the temperature and density of the nebula; as a consequence, the nebulae emit radio waves with a balanced proportion of long and short wavelengths which is characteristic of thermal radio sources.

Radio supernova remnants, however, are non-thermal. The explosion of a supernova puffs off a shell, or spherical bubble. The shell catches up the interstellar magnetic field, and also contains the star's own magnetic field, which is trapped in the outflow and compressed. Electrons in interstellar space can be accelerated by bouncing off the magnetic field embedded in the heavy shell. Their speeds become comparable to the speed of light and, in contrast to thermal electrons, they are said to be relativistic (because the theory of relativity is needed to describe them). As an alternative mechanism the relativistic electrons can be generated by a central pulsar in the supernova remnant. Wherever they come from, the electrons spiral around, twist in and bounce off the magnetic fields in the supernova remnant. These are changes in motion and therefore these electrons also emit radiation, so the supernova remnant gives off radio waves. The balance of long and short wavelengths emitted by the supernova remnant depends on the magnetic field as well as the motions of the electrons. While not unique, the spectrum of a supernova remnant is sufficiently different from the spectrum of a thermal radio sources to be distinguished as non-thermal. In cases of doubt, the clinching argument is to look for the faint radio spectral lines emitted

by hydrogen recombining in the nebula – these are found only in thermal radio sources.

Because radio astronomers can see supernova remnants across the Galaxy they have the opportunity to map all the supernovae which have occurred in the last 10 000 years or so, the length of time for which a radio supernova remnant is identifiable. In fact, radio astronomers have been able to produce the only generally applicable method for estimating the distances of supernova remnants. It is the method of *H I distances*.

The galactic plane is full of cold hydrogen (H I – pronounced 'H-one') clouds which orbit around the Galaxy. If seen against the cold of deep space these clouds can be seen to emit 21 cm radio waves; but if seen in silhouette against a bright supernova remnant they absorb 21 cm radiation. The exact wavelength at which the radiation is absorbed depends on the velocity of

Table 2. *Some supernova remnants*

Name(s)	Supernova date	Distance (l.y.)	Description[a]	Supernova type
Cassiopeia A	AD 1658±3	9 100	R O X S	II
Kepler's SN	AD 1604	26 000	R O X H S	I
Tycho's SN, 3C 10	AD 1572	10 000	R O X H S	I
Sagittarius A East	AD 1300±	30 000	R S	
3C 50	AD 1181	28 000	R O X H C D	
Crab, Taurus A, 3C 144, M1, NGC 1952	AD 1054	6 500	R O X H P D	II
P 1459–41	AD 1006	4 200	R O X H S	
G 292+1.8	AD 1000±	12 000	R O X S	II
MSH 15–52	AD 320?	15 000	R O X P S	
Puppis A	2000 BC	7 200	R O X S	
Vela	8000 BC	1 600	R O X P S	
CTB 109	10 000 BC	10 000	R O X C S	
Cygnus Loop, Veil Nebula	20 000 BC	2 600	R O X S	
W50, SS433	30 000 BC	15 000	R O X C S	
IC 443	60 000 BC	5 000	R O X S	

[a] Key to description: R, radio remnant; O, optical remnant; P, pulsar known; X, X-ray emitter, H, historical supernova; S, shell; D, disc; C, compact stellar remnant.

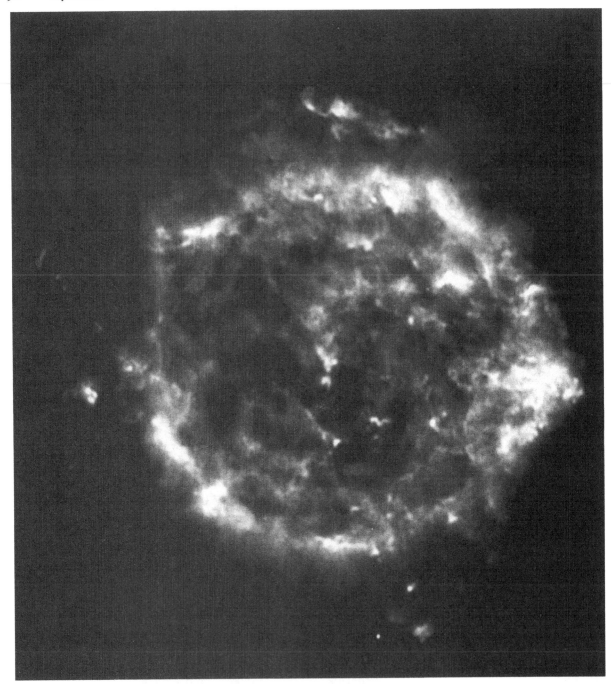

FIG. 45. *Cas A at 5 GHz in 1978. With radio telescopes, the Cassiopeia A supernova remnant shows as a complete spherical shell like Tycho's remnant. However, it has numerous clumpy knots like the optical image of Cas A. This radiograph is by Steve Gull and the staff of the 5-km Cambridge radio telescope and has been sharped by the Maximum Entropy Method to yield the finest possible detail.*

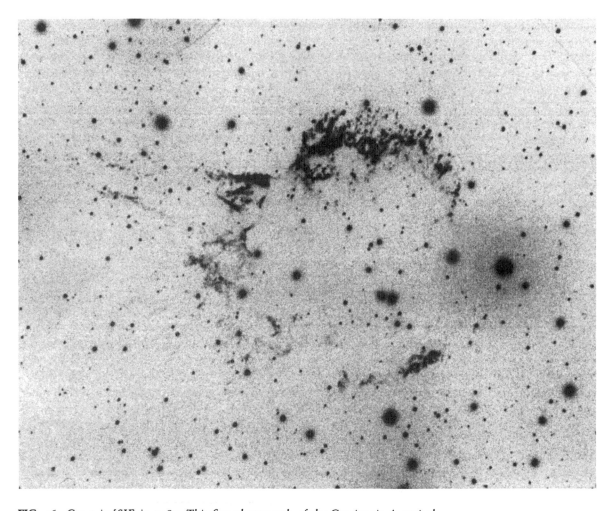

FIG. 46. *Cas* A *in* [SII] *in 1983. This fine photograph of the Cassiopeia A optical supernova remnant shows a generally circular, but broken, shell of fragments of the exploding star. It was made under excellent conditions by Sidney van den Bergh with the Palomar 200-inch telescope in 1983. He used a coloured filter in the telescope to emphasize the sulphur spectral lines [SII], of the supernova remnant and de-emphasize the images of the numerous Milky Way stars. The negative photograph shows stars as hard black circles; supernova fragments are irregular patches, the fainter ones grey rather than black.*

the cold interstellar cloud, and the velocity in turn depends on the position of the cloud in our Galaxy. So, if a radio astronomer looks at a radio supernova remnant and at a region of space off to one side so that he can identify which clouds he can see in absorption against the H I cloud and which are beyond it, he can gain an idea of the position of the remnant. For instance, if in the nearby area he sees 21 cm radiation at three different velocities corresponding to clouds at distances of 10 000, 15 000 and 25 000 light years, whereas in the supernova remnant he sees the absorption of 21 cm radiation only at 10 000 and 15 000 light years, then this places the distance of the remnant at between 15 000 and 25 000 light years.

Thus radio astronomers distinguish, define and map radio supernova remnants, most of which were left by supernovae seen, if bright enough, by the uncomprehending and apprehensive eyes of Stone Age humans. Behind some of the 135 radio supernova remnants lie fascinating stories of astronomical detective work by their modern descendants.

Cassiopeia A

The most intense radio source in the sky is in fact a supernova remnant. In the constellation Cassiopeia, it is known as Cassiopeia A, Cas A for short. It was even visible but unrecognized on G. Reber's first maps of the sky in radio waves, and it was rediscovered in 1948 by Martin Ryle and F. Graham Smith who were at Cambridge. Smith determined its position accurately and, in 1954, Baade and Minkowski discovered the optical counterpart of Cas A. They found an almost complete shell of knots and filaments approximately 4 arc minutes across.

Some of these knots and filaments were shown by Baade and Minkowski to be expanding from a central position. No star is detected there, so that there is no neutron star visible. The knots appear to come and go, with only one third of the knots being visible for more than 12 years. They are probably produced by part of the supernova ploughing through stationary interstellar cloud banks. As they strike the cloud banks, atoms in the knots become what physicists term 'excited'. Electrons in atoms of oxygen, sulphur and argon are moved to high energy states. When they drop back to their original low energy states, they emit spectral lines of these elements.

Some very fast-moving knots are speeding away from the central part of Cassiopeia A at a speed of 9500 km/s. The ejection velocity and momentum in the knots is so high that collisions do not appear to have slowed the knots significantly since they were first ejected. This assumption enables the date of the Cas A supernova to be calculated. From the expansion velocity, Sidney van den Bergh and K. W. Kamper have deduced that the supernova which produced Cassiopeia A took place within 3 years of 1658.

The strange thing is that no supernova is recorded in Chinese, Korean or Japanese chronicles at this date, which suggests that when this Cassiopeia supernova occurred, it was fairly faint. Its distance is comparable with that of Tycho's supernova, namely, just over 9000 light years. It may have appeared fainter than Tycho's supernova because it occurred in a very dusty region of the Galaxy and its light was obscured. There is some indication that this is the case and that the supernova at its brightest was just about visible to the naked eye.

Flamsteed's supernova

Although the Chinese and Japanese sources do not record a supernova at the appropriate time, John Flamsteed, the first English Astronomer Royal was doing his job so carefully at the Royal Observatory in Greenwich that he seems to have recorded the supernova even though it was so faint that he assumed that it was one of the ordinary stars of Cassiopeia never before noticed or catalogued. After the founding of the Royal Observatory in 1667 Flamsteed had begun making very careful observations of the positions of celestial objects. The only instrument available to him in the early years was an astronomical sextant. He recognized that its accuracy was poor and he longed for better equipment so that he could achieve greater accuracy in taking positions.

Even with his sextant, however, Flamsteed did achieve surprising accuracy. He considered that he was able to do much better than the great observational astronomer Tycho Brahe whose observations were accurate to 3 or 4 minutes of arc. Flamsteed's care in observing is apparent from the discovery by an editor of his work that out of 2935 stars observed, only 4 have not been identified in subsequent observations. The probability, then, is that when Flamsteed said he saw something, he did in fact see something. On 1680 August 16 he measured and recorded in his notebook a star in Cassiopeia, calling it simply *supra tau* or *the star above tau Cassiopeiae*. In Flamsteed's published catalogue of 1725 the star was listed as 3 Cassiopeiae. It has puzzled editors ever since because there is no star in that position. Caroline Herschel, when editing Flamsteed's works, confused his observations of 3 Cassiopeiae (3 Cas) with an observation of another completely different nearby star called AR Cassiopeiae and dropped number 3 from her *Index . . . to Mr Flamsteed's Observations*. Francis Baily sorted out

the confusion in his edition of Flamsteed's *British Catalogue of Stars*, but could not explain how Flamsteed had observed a star which was not there. He remarked, with characteristic Victorian understatement, that the whole affair was 'singular'.

Was the star which Flamsteed saw once and never again the supernova which left Cas A? There are two uncertainties. In the first place van den Bergh and Kamper estimated a date for the explosion of AD 1658±3 years. If that calculation, based on an assumption of zero deceleration, is accurate then Flamsteed would not have been able to see the supernova in 1680. If we none the less continue to assert that he did, then the estimated date is wrong by about 20 years (6% of the age of the remnant). In fact this is quite possible, and not excluded by the observations and measurements by van den Bergh and Kamper. The filaments ejected by the supernova lose momentum as they gather up stationary material from the interstellar medium, just as a ball slows when it gathers a gloved hand. When a filament has gathered stationary material equal to its own mass, its speed has been halved, and its age would seem twice as long as it really was. A discrepancy of 6% in the age of the supernova can be accounted for by the supernova filaments sweeping up from interstellar space about 6% of the mass of the supernova. This is not at all impossible, since the explosion took place in a dusty part of the Galaxy where the interstellar material is dense and a hollow the size of Cas A could easily have contained several solar masses of interstellar hydrogen.

A second problem is the position. Although Flamsteed's position for 3 Cas is very near to the position of Cas A, he missed the exact position by 12.1 arc minutes in right ascension and 8.6 in declination. The total is about half the diameter of the Moon and is by no means insignificant. William B. Ashworth has recalculated the position of 3 Cas by using modern positions of the stars with respect to which Flamsteed measured 3 Cas. This reduces the discrepancy somewhat. But for the residue, about 13 arc minutes in total, we might look to Flamsteed's measuring error. Flamsteed boasted of his superior technique compared to those of Tycho Brahe and Johannes Hevelius, and, indeed, generally his errors were 1–2 arc minutes – about half theirs. However, Ashworth quotes three stars in Flamsteed's *British Catalogue* which are off by 9–25 arc minutes. The error in the position of 3 Cas, if it is indeed Cas A, is not out of the question. Possibly, Flamsteed did indeed observe the Cas A supernova, and it was just visible to the naked eye.

The radio picture

Very detailed radio maps have repeatedly been made of Cassiopeia A; they show a complete, bright circular shell with numerous clumpy knots which, like the fast-moving optical knots, come and go. The radio astronomers who have made the maps have attempted to determine the outward expansion of Cas A, like the optical astronomers, but the radio astronomers have not been able to agree amongst themselves as to the whole story. In fact, one memory of the conference on supernova remnants held on the beautiful island of San Georgio in Venice in the late summer of 1982 was of the contrast between the peaceful summer evenings as the sun set over the Venetian Lagoon and the tense atmosphere inside the conference room as the rival groups of radio astronomers presented their contradictory results.

Altogether there have been four attempts to measure the radio expansion of Cas A. The first used maps made 5 years apart with the

Cambridge 1-Mile and 5-km Radio Telescopes. A. R. Bell looked at 30 compact knots and found (in 1977) evidence for expansion, but the measurements showed a good deal of scatter. Although the overall motion was outwards, some knots appeared to be moving circumferentially around the supernova remnant. It is hard to think how any mechanism can make the knots do this, and measurements which show this must be interpreted cautiously. Shortly after Bell's measurements, Dickel and Greisen used a National Radio Astronomy Observatory (NRAO) telescope and found no evidence for systematic outward expansion of Cas A, although there were random motions. Insofar as there was a discrepancy between the two results – some astronomers concluded from the published papers that the results were not really inconsistent – the discrepancy could be explained by the way in which the observations were made. The Cambridge study used two different telescopes, and it would be hard to relate their pictures in fine detail. Slight differences in appearance of the knots could be misinterpreted as motions. The motions would appear random. The NRAO study, on the other hand, used the same telescope to obtain its two pictures at different epochs but the observations were incomplete; the US radio astronomers made assumptions on how this could be accounted for and, maybe, these assumptions gave rise to spurious results.

The Cambridge group repeated their work with two images of Cas A obtained in 1974 and 1978 with the same telescope, the Cambridge 5-km Radio Telescope. R. J. Tuffs identified 342 radio peaks on the maps and showed that they systematically moved outwards from the expansion centre at speeds 3 times slower than those of the optical knots. The differences in speed are not surprising as the optical knots

FIG. 47. *The Cas A supernova? Tycho's supernova of 1572 is the most prominent object on this enlargement of the Cassiopeia plate of Bevis' celestial atlas. A chain of four stars grouped in two pairs runs to the upper left corner, continuing the line of Cassiopeia's upper arm. The third star from the left is 3 Cassiopeia, as observed by Flamsteed in 1680; but it no longer exists. It lies close to the radio and optical supernova remnant Cas A – in fact, on the scale of this atlas, the radio supernova remnant lies within the drawn image of the star. It seems that Flamsteed's observation of 3 Cas is the only observation of the supernova. From the archives of the Royal Greenwich Observatory.*

represent fragments of the exploding star, whereas the radio peaks represent material disturbed by them – there would, similarly, be a difference in speed between a torpedo and the water following in its wake. There was a considerable random pattern in the motions of the Cas A radio peaks.

'The episode would thus appear to be neatly concluded and the dispute settled', wrote Richard Strom in a masterly and diplomatic summary of the affair, 'were it not for the observations recently carried out with the VLA'. The VLA is the Very Large Array of radio telescopes in the New Mexico desert near Socorro. Combining the largest number of radio telescopes dispersed over the largest distance enables radio astronomers working with the VLA to see the finest details in a radio source; P. E. Angerhofer and R. Perley have begun to use the VLA to monitor Cas A. In their first results, after 2 years of observation, they have measured the displacements of 120 radio peaks in the supernova remnant; they observe no overall expansion! This is in qualitative agreement with the results of J. Dickel and E. Greisen but contrary to those of the two Cambridge groups. It is probable that the dispute will not be quickly or quietly settled.

Sagittarius A

Sagittarius A was discovered in the 1950s; it is a complicated radio source in the direction of the centre of our Galaxy and it has several parts which are superimposed on one another. In 1972 it was found that Sagittarius A had two halves, which are named Sagittarius A East and Sagittarius A West. Sgr A West has a thermal radio spectrum: it is a small nebula. Because there is so much dust along the line of sight to the galactic centre, no light from the nebula penetrates to the Earth. However, new electronic detectors called Charge Coupled Devices (CCDs) are sensitive to the infrared, which has a longer

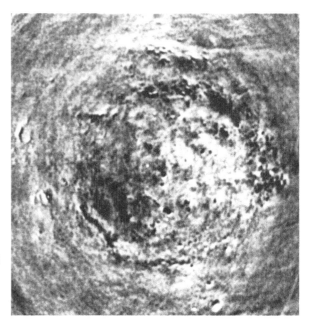

FIG. 48. *Cas A expansion. Two images of Cas A at 5 GHz radio frequency are combined in this image, one obtained in late 1974 and presented black on white, the second in 1978 and presented white on black. Individual patches of the remnant show a generally outward motion. Compare this radio picture of Cas A by Richard Tuffs and the staff of the 5-km telescope with the optical picture of the Crab by Virginia Trimble in Fig. 34.*

wavelength than light and penetrates the dust more easily. With CCDs it has proved possible to record images of two small infrared nebulae at the position of Sgr A West. Near to these two nebulae is a compact, point-like radio source which radio astronomers believe to be a mini-quasar at the very centre of our Galaxy, about 30 000 light years from the Sun.

Sagittarius A East, on the other hand, has a non-thermal radio spectrum, and this suggested that it might be a supernova remnant. In the early 1970s, two teams showed that Sgr A East was arc-shaped, but more than that could not be said. The whole of Sgr A is small and the source is in the southern sky, so radio interferometers in the northern hemisphere find it difficult to study; radio interferometers in the southern hemisphere, while able to look straight up at the source, do not have the resolution (ability to discern detail) to map it well. The nature of Sgr A East was therefore in doubt until, in 1982, Ron Eckers, J. van Gorkom, Miller Goss and Ulrich Schwarz used the Very Large Array (VLA) of radio telescopes at Socorro, New Mexico, to map it. They discovered that Sgr A East was part of a complete shell which overlapped Sgr A West, and was clearly a supernova remnant. It is about 30 light years in diameter and is the third brightest radio supernova remnant in our Galaxy, after Cas A and the Crab Nebula.

No-one is sure whether Sgr A East is actually at the galactic centre or just nearby (say, within 1000 light years of the centre, inside the nuclear bulge of our Galaxy). Its age is in the range 140–440 years (judging by its size). However, if its parent supernova exploded inside a very dense cloud (of which several examples are known in this direction), the surrounding material would hold in the explosion and the age of the remnant could be much greater.

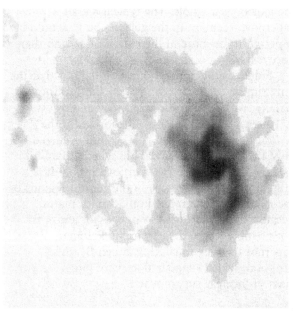

FIG. 49. *Sagittarius A. The Sagittarius A radio source shows as an overall elliptical shell, on which is superimposed (right) a small, brighter structure (black on this negative representation). The bright structure is very complex; it is called Sagittarius A West and is a distant nebula at the galactic centre. The larger elliptical shell, showing grey, is called Sagittarius A East, and is a supernova remnant a few hundred years old, near the galactic centre. Picture is courtesy of Miller Goss.*

It is surprising that a supernova has occurred in the nuclear bulge as recently as a few hundred years ago, since although the stars are densely packed in this region and it is the brightest part of the Galaxy, there are not actually many stars there in total – the nuclear bulge of a typical spiral galaxy contains 1% of its light and 0.1% of its mass. The mean time between supernovae in this part of a galaxy should therefore be 100 to 1000 times the mean time between supernovae in the galaxy as a whole. The typical age of a supernova remnant in the galactic bulge would be expected to be more in the region of 10 000 than 100 years.

Sidney van den Bergh has pointed out that the only supernova ever seen in M31, the Andromeda Galaxy, was S Andromedae, discovered by E. Hartwig (see Chapter 4). It too was in the central bulge of its spiral galaxy, and occurred only 100 years ago. Maybe the centres of spiral galaxies are particularly supernova-prone. It is difficult to see supernovae in the central regions of the galaxies photographed on Schmidt plates, because they are heavily overexposed, and it is not possible to use data from the Palomar supernova survey to check van den Bergh's proposal, which suggests that there might be a class of 'nuclear supernovae'.

Puppis A

Just as Cassiopeia A is the brightest radio source in the constellation of Cassiopeia and Taurus A the brightest in Taurus, so Puppis A is the brightest radio source in the southern constellation of Puppis and, like them, it is a supernova remnant. In radio maps it has the usual shell-like shape characteristic of supernova remnants, X-ray images taken with the Einstein Satellite show the same picture. Optical photographs of the area show curlicues, wisps and filaments of nebulosity. From the size and brightness of Puppis A, radio astronomers deduce that it is about 4000 years old. It is therefore not surprising there are no records of historical supernovae which could be associated with it – there were then no literate civilizations in the southern hemisphere to pass on to us their vision of the Puppis A supernova.

Astronomers regard Puppis A as unusual because spectra of the filaments show that they

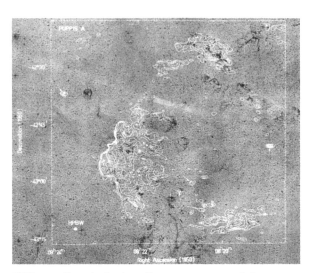

FIG. 50. *Puppis A. A radio contour map of the supernova remnant Puppis A is superimposed on a photograph of the same region of sky. The radio remnant shows an incomplete hollow shell and is superimposed on a motley collection of curlicues and filaments of nebula, which represent the remains of the exploding star in collision with its circumstellar envelope. Courtesy of Miller Goss.*

are rich in nitrogen. This area of the Galaxy is probably not different from any other, so that the nitrogen overabundance is not likely to be a characteristic of the interstellar medium near to Puppis A. Instead, some astronomers interpret Puppis A as an example of a star exploding in its own circumstellar shell.

Their story goes like this. The Puppis A supernova remnant was the result of a massive star exploding. In the course of its life and evolution, the star had lost matter in a stellar wind. Our own Sun loses matter in a solar wind, which is pushed out by radiation pressure and flows past the Earth's orbit, beyond, to the edge of the solar system. In bright stars, the winds can be much more effective than in our Sun, and a substantial fraction of the star's mass can be lost in a circumstellar envelope. The star has been shining by synthesizing heavy elements from hydrogen nuclei. The nuclear synthesis takes place within the star and, as the outer layers of the star peel off, the wind carries outwards the nitrogen-rich material. Eventually, the star explodes as a supernova. Instead of showing itself to us by means of interstellar material which just happens to be near to the star, the supernova remnant is composed of the circumstellar material discarded by the star itself in a previous phase of its life.

X-rays from supernova remnants

Because the Crab Nebula was the first supernova remnant to be seen in X-rays, and because it had been well established that its X-rays were from synchrotron radiation, the first attempts to understand the spectra of other X-ray supernova remnants were in terms of the synchrotron theory, but a totally different mechanism is at work in most X-ray supernova remnants.

The explosion of a supernova takes place in interstellar space but not in a perfect vacuum – it occurs in the interstellar medium, and, rarefied though that is, its density is not negligible. We have already seen that the cumulative effects of dust in interstellar space, although there is on average only one dust grain per cathedral-sized volume of space, can mask the light from supernovae which occur on the other side of our Galaxy. Similarly, although interstellar space contains a gas of superlative tenuousness – with a density of one atom per cubic centimetre the gas is much more rarefied than what on Earth is classed as a vacuum – nonetheless a supernova explosion occurs over such a large volume that the remnant quite quickly sweeps up interstellar material of a mass comparable with that of the

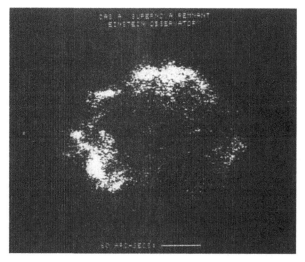

FIG. 51. *X-rays from Cas A. The Einstein X-ray satellite image of the Cas A supernova remnant shows a complete shell reminiscent of the radio and optical pictures (Figs. 45 and 46). The X-ray picture of Cas A is unlike that of the Crab in that it is an open shell, whereas the Crab has a filled centre.*

star which exploded. Even the tenuous gas and dust of interstellar space therefore have an effect upon the evolution of a supernova remnant, clinging to the exploding fragments and dragging them to a stop.

The dust swept up by the Crab Nebula was recorded in 1984 by a group of astronomers using IRAS, the American–British–Dutch collaborative InfraRed Astronomy Satellite. They observed the Crab at long infrared wavelengths in the region of 60–100 µm. The main radiation which they saw was the synchrotron radiation, the infrared observations bridging the gap between the optical and radio parts of the spectrum. Superimposed on this radiation was additional infrared emission – a so-called infrared excess. It came from interstellar dust swept up by the expanding supernova remnant, and heated to about 70 degrees Kelvin by absorbing the ultraviolet and X-radiation from the Crab. It amounts in quantity to only a few thousandths of a solar mass of dusty material.

The effect on the swept up gas is even more dramatic. Within a day or two of the supernova explosion, the moving pieces of star have grown to an object the diameter of the star's solar system and are driving into the gas of the interstellar space beyond its reaches, like a piston compressing fuel in a diesel engine. Just as the diesel vapour is heated (to such a temperature that it ignites and drives the engine, in fact), so the interstellar gas is heated. The piston of the supernova explosion heats the interstellar gas to temperatures of millions of degrees and, at such temperatures, the gas emits X-rays.

The speed of the expansion of the supernova into the gas is many thousands of kilometres per second and this is much larger than the speed of the atoms of the gas. The interstellar atoms do not have enough time to get out of the way of the oncoming pieces of star. They are caught up in a *shock*, like snow in a snowplough, and X-ray images of supernovae show all the shock-heated material in the interstellar gas around the supernova. Some of the shock-heated material emits light and we can photograph optical filaments. Except in the case of the Crab, and maybe other Crab-like supernova remnants, this is the route by which all optical supernova remnants are made visible. (The Crab's filaments are energized by the ultraviolet light from the synchrotron radiation emitted from within.) The shock-heated material is in the shape of a hollow sphere. Our line of sight to the supernova remnant passes directly through the thickness of the front and back sides, when we look through the centre of the sphere, but when we look at the edges of the remnant, our line of sight grazes into the thickness of the sphere and passes through a

longer length of the shock-heated material. At the edges of the remnant, therefore, we see more X-rays than come from the middle. In astronomical jargon, the edge of an object is called its *limb* and this phenomenon is called *limb-brightening*. The limb-brightened X-ray images of typical supernova remnants, like Cas A and Tycho's supernova remnant, have a typical shell-like appearance forming complete rings of X-rays.

Using the Einstein satellite's High Resolution Imager, Fred Seward and his colleagues at the Harvard–Smithsonian Center for Astrophysics have identified the bright rim of the Tycho supernova remnant and the clumpy material which appears as patches all over the remnant's face as the ejecta from the exploding star. They estimate that there is a total of 1.9 solar masses of ejecta, but the exact value depends on the distance that they assume for Tycho's supernova, and this is quite uncertain (8000–15 000 light years). They can see a diffuse region around the supernova remnant and identify this with the heated shock in the interstellar medium. They calculate that the remnant has swept up about its own mass of interstellar medium.

The spectra of the X-rays from the shock-heated material show some spectral lines. As in optical spectral lines, the X-ray lines represent transitions of electrons in atoms and ions in heated material. The X-ray photons which are emitted in these processes are, however, 1000 times more energetic than optical photons, and the electron movements which give rise to them occur deep within the atoms where the electrons are tightly bound, rather than in the atoms' more loosely bound outer regions. Close to the atomic nuclei, the electrons in atoms are much more energetic, so changes in their energy as they move from one level to another are likely to be large. Atoms and ions do not usually suffer

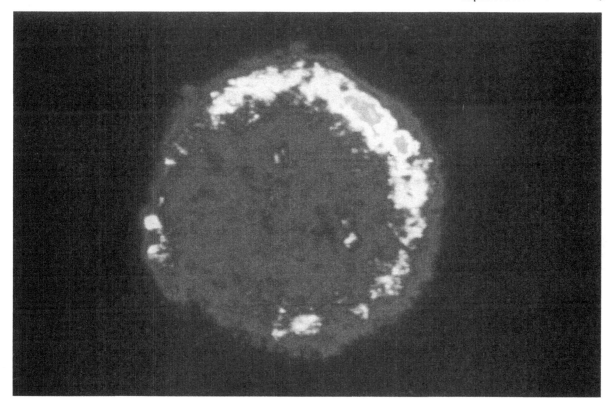

FIG. 52. *Tycho's supernova remnant in X-rays. The Einstein Satellite image of Tycho's supernova remnant shows a hollow shell structure. The outer rims of the shell and the brighter patches on the front and rear faces of the shell, projected along the line of sight into the middle of the shell, are ejected material from the exploding star. Visible in this picture as a halo around the bright rim, a faint outer extension to the remnant represents the interstellar medium, which has been shocked by the out-rushing ejecta and heated to temperatures of millions of degrees. This is sufficient to emit faint X-rays. Courtesy of Fred Seward, Harvard-Smithsonian Center for Astrophysics.*

transitions like these unless they are strongly heated or exposed to energetic X-rays from outside – but these are just the conditions in supernova remnants! The X-ray line spectra which thus occur provide a new way of looking at the constituents of supernova remnants – i.e. the interstellar medium – and of deducing more about the composition of our Galaxy.

Plerionic supernova remnants

We have already seen that the Crab Nebula is, optically, a hollow shell of filaments filled by synchrotron radiation. The synchrotron radiation gives rise to milky white light, which mingles with the light from the filaments, and also to the Crab Nebula's radio and X-ray images. The radio and X-ray synchrotron radiation from within the filled sphere far outshines any radiation from the shell; in contrast to most supernova remnants, the Crab in X-rays and radio is not limb-brightened – it is brightest in its centre and is therefore *limb-darkened*. The reason is that our line of sight through the centre of the sphere containing the synchrotron radiation passes across the supernova remnant's diameter, but only just touches the sphere when passing near its edge.

The Crab is the archetype of the so-called filled-centre supernova remnants, or *plerions*, a name coined by Kurt Weiler. Weiler has identified about half a dozen plerions (with numerous

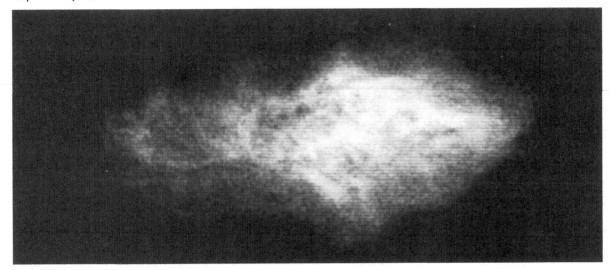

FIG. 53. *Plerion. This radio graph of 3C58, the remnant of the supernova of 1181, has been made from observations with the Mullard Radio Astronomy Observatory's 5-km radio telescope by Dave Green and Steve Gull. Like the Crab Nebula, it has a filled appearance, rather than the shell-like look of other supernova remnants. Four times more distant than the Crab, it is almost six times its size, although of comparable age. This is because it has exploded at six times the average speed of the Crab supernova. Green and Gull's picture has been computer-sharpened by the Maximum Entropy Method and reveals a filamentary structure within the radio synchrotron emission, just like Elizabeth Swinbank's picture of the Crab (Fig. 38).*

possible additions and special cases). The implication of the filled-centre, limb-darkened shape is that within such remnants lies an active source of relativistic electrons which give rise to the synchrotron radiation – and it is natural to suspect that each plerion contains a pulsar. In spite of the most diligent searches, however, pulsars have been identified in only two plerions: the Crab itself and, most recently, a supernova remnant in the Large Magellanic Cloud called 0540–69.3. Never willing to abandon an inspired guess without a struggle, astronomers have turned this partial defeat into a victory by arguing that this is just what you would expect. Only about one pulsar in ten or so can be detected, because they have beams which are not visible from all angles. Think of a lighthouse, whose beams are visible from a distant aeroplane as it approaches the sea shore. As the aeroplane flies over the lighthouse, the pilot, looking down, sees no beams – he is looking down the pole of rotation of the lighthouse mechanism and the beams are at right angles to this pole. The pilot can, however, see the rocks around the lighthouse, as the beams

illuminate them. Similarly, pulsars whose poles of rotation point to the Earth cannot be seen, although we may be able to see the synchrotron radiation from the nebula surrounding the pulsar. Thus most astronomers are willing to believe that the plerions contain pulsars, although in only two can a pulsar, as such, be detected.

FIG. 54. *Gum Nebula. The constellations of Vela, Pyxis, Puppis and Carina form a rich region of the Milky Way, because it is here that we on Earth look tangentially along a spiral arm of our Galaxy, rather than across one. We see spiral arm material – nebulae, dark clouds, masses of stars – which extends a long way into the distance, along the line of sight. Nearby to us, however (only 1500 light years away) is the Vela supernova remnant, just visible as filaments of nebulosity right of centre of this montage of red-sensitive photographs of the region. At the top right edge of the montage, at the top centre, and in the bottom third of the region shown, are parts of the Gum Nebula. It entirely surrounds the Vela supernova remnant and is approximately centred on it. The speculation is that the nebula was illuminated by the wave of light spreading out from the Vela supernova explosion of about 10 000 years ago; the gas ionized by the explosion has not yet had time all to recombine.*

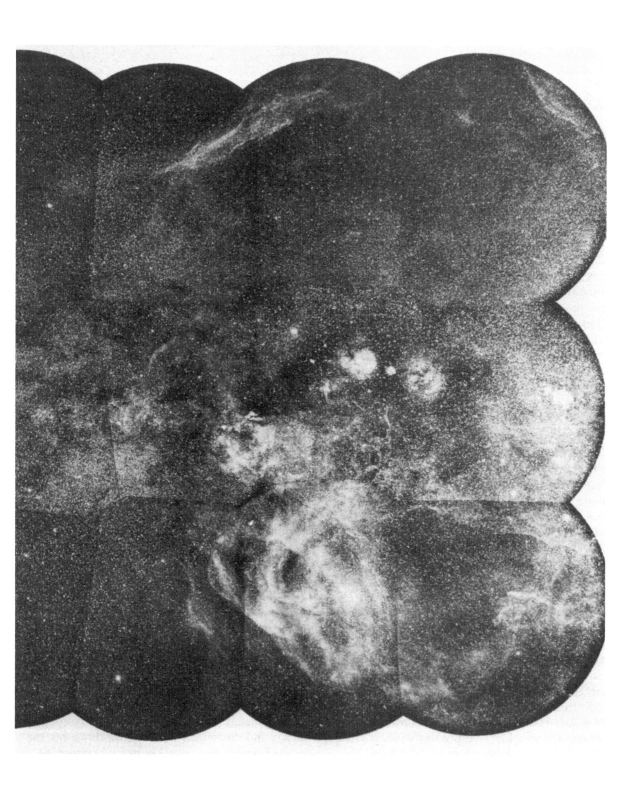

The clearest plerion, apart from the Crab, is 3C58, identified by Richard Stephenson as the remnant of the supernova of AD 1181 (see Chapter 1). It has a bright elliptical shape and is brighter in the middle than at the edges. A few faint optical filaments which look like the Crab's have been discovered at its position, but the region is heavily obscured and optical observations are not much help in this case. 3C58 was detected as an X-ray source by R. H. Becker using the Einstein Satellite: it is filled in the centre and has a synchrotron spectrum. Centrally in the diffuse X-ray emission from the supernova remnant lies a faint compact X-ray star which is presumably the stellar remnant of the supernova of AD 1181, but the mechanism which produces its X-ray emission is not clear.

The Gum Nebula and the Vela supernova remnant

The largest nebula in the sky is called the Gum Nebula. It was discovered by a young Australian astronomer, Colin Gum, whose career was cut short by a fatal skiing accident at the age of 36. For his PhD thesis, Gum had photographed the whole of the Milky Way as it is visible in the southern hemisphere. He used a red filter so that the photographic plate would pick up predominantly the deep-red spectral line emitted by hydrogen, H-alpha. Because of the predominance of hydrogen in the interstellar medium, H-alpha is usually the strongest spectral line emitted by nebulae. At the same time, the light of other objects in the heavens, including the glow of our own atmosphere, is fairly weak at the red end of the spectrum. By preventing all but red light from reaching the photograph, it is possible to isolate H-alpha-emitting objects from the others, and give longer exposures without causing white light from other objects to flood the

photograph. Gum made a mosaic of several of his photographs of the constellations Vela and Puppis, and the nebula which now bears his name at once became apparent. It has a diameter of about 30 degrees on the sky. This means that if a person's eyes were sensitive enough to perceive the faint H-alpha emissions from the Gum Nebula, it would fill half the area of the sky that one eye can see at a time.

Within the Gum Nebula lie the two stars Zeta Puppis and Gamma Velorum. Gum thought that these two stars were emitting enough ultraviolet light to be able to separate, or *ionize,* any hydrogen atoms in the vicinity into electrons and protons. The result would be that when these electrons and protons recombined into hydrogen atoms, they would emit the H-alpha light that he saw coming from the nebula. The recombined atoms would almost immediately be re-ionized by the two stars' ultraviolet light, to recombine again in a repetitive cycle; the nebula is constantly regenerated. The two stars are 1500 light years away from the Earth, while the Gum Nebula is approximately 1000 light years in radius, so that the Sun lies close to the nearer edge, not quite within the Gum Nebula, but not far away.

In 1971, a group of four astronomers, John Brandt, Theodore Stecher, Steve Maran and David Crawford, calculated that Gamma Velorum and Zeta Puppis did not produce enough energy to be able to split apart the large number of hydrogen atoms within such a large volume of space. They said instead that the hydrogen atoms of the Gum Nebula had originally been split apart, not by the steady radiation from the stars within it, but by a burst of radiation from the explosion of a supernova some 10000 years ago. The ionized gas will ultimately all recombine to make hydrogen atoms and the nebula will cease to shine, but the recombination process takes many hundreds of

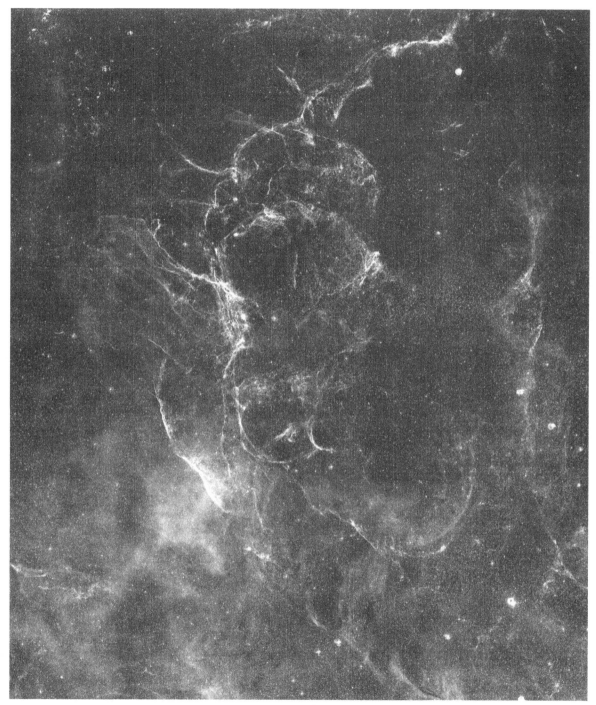

FIG. 55. *Vela supernova remnant. The brighter half of the filamentary shell of the Vela supernova remnant fills most of this photograph of the picture. It was taken in the red light of H-alpha by John Meaburn and Ken Elliott, using the 48-inch Schmidt Telescope at Coonabarabran, Australia. The Vela pulsar lies at centre right, though far too faint to show. The milky diffuse patches, most prominent at lower left are parts of the Gum Nebula. The motley collection of small fragments in the upper left is Puppis A.* © *Royal Observatory, Edinburgh.*

thousands of years because the distance between the electrons and protons is so vast that a wandering electron seldom passes close enough to a proton to be grabbed and made into a hydrogen atom.

According to this interpretation, the Gum Nebula is not a nebula in the usual sense since the stars present are insufficient to keep the nebula constantly ionized. It is a new kind of nebula created suddenly by a supernova explosion and gradually returning to its normal state, not being renewed. The Gum Nebula is a fossil left behind by an event long past.

One must admit that many astronomers are unhappy with the interpretation of the Gum Nebula as the fossil of a relatively recent supernova explosion. Some argue that the Gum Nebula may be a remnant from a supernova that exploded a million years or more ago. They argue that its size has been overestimated and that its two stars, Gamma Velorum and Zeta Puppis are indeed capable of ionizing the hydrogen within it

in order for it to emit H-alpha light.

Is there evidence that a supernova did occur somewhere near the centre of the Gum Nebula about 10 000 years ago? The answer is yes. In the centre of the Gum Nebula is the smaller nebula known as the Vela supernova remnant. This supernova remnant was discovered by Douglas Milne in 1968 using a radio telescope, and we first came across it in Chapter 6, in connection with the discovery of pulsars. Photographs of the area show a filamentary nebula looking just like other well-known supernova remnants, such as the much-photographed Veil Nebula in the northern sky. A little off-centre from the Vela supernova remnant lies the Vela pulsar. The Vela pulsar's period is just less than $1/10$ second, and is changing at such a rate that it doubles its period every 11 000 years. This can be taken as a measure of the age of the pulsar, which is consistent with the characteristics of the Vela supernova remnant and with estimates of its age from other evidence.

8

Types of supernovae

Astronomers are convinced of the connection between supernovae, pulsars, and supernova remnants, but they are also aware that a supernova does not appear from nothing. Which stars become supernovae?

Two kinds of supernovae

It turns out that there are at least two kinds of stars which produce supernovae: only one kind has been confidently identified. This answer emerged when the supernovae themselves were first studied; it was Fritz Zwicky who organized the effort required to determine it. When he first set up the Palomar Supernova Search, Zwicky arranged a cooperative effort to follow the discovery of a supernova by detailed study. Walter Baade would measure the light curves and Rudolph Minkowski and Milton Humason would obtain spectra of the supernovae using the largest telescopes then available, the 60-inch and 100-inch Mount Wilson telescopes. Like other professionally used astronomical telescopes, these are scheduled in advance, so that a particular night is preassigned to a specific astronomer to work on a project whose value he has justified to a committee, in competition with other observers who also wish to have the use of the telescope. Because present-day large telescopes are oversubscribed by a factor of two or three, the competition is fierce. There is not enough time on the telescope to satisfy all astronomers.

By their character supernovae are unpredictable, therefore one cannot preassign time on telescopes for their study. Zwicky persuaded the Mount Wilson Observatory Director, George Hale, to set up an override on observing time on the big telescopes at Mount Wilson such that the scheduled astronomer had to yield time for Baade or Minkowski to observe supernovae bright enough to be worth studying. An informal system

now operates at Palomar and at other large telescopes; the scheduled astronomer is not obliged to give way, except insofar as he wishes to continue to have good relations with his colleagues, or owes them a favour.

From studies of spectra of the first supernovae discovered in Zwicky's Supernova Search, Rudolph Minkowski found in 1940 that most supernovae discovered (three quarters) are of a single type, which he called Type I. The remainder were different from Type I, though not all alike, and they are known as Type II. Broadly speaking, Type II spectra showed hydrogen lines and Type I didn't. (Zwicky has been able to recognize further Types III, IV and V, but these types consist of isolated examples and not all astronomers are convinced that they are distinct kinds of supernovae.)

The Type I supernova has an instantly recognizable light curve, consisting first of the sudden spectacular brightening, a quick decline in less than 1 month, and then a slower fade-off. In the case of the supernova of 1937 in the galaxy IC 4182, the brightest supernova of this century, astronomers followed this fade-off for more than 600 days during which time the supernova faded without showing signs of stopping its decline. This supernova light curve is considered the prototype of Type I. The slow fade-off always occurs at the same rate in Type Is, such that the light output halves every 60 days.

Wondering whether the supernovae seen in our Galaxy were of Type I or not, Walter Baade plotted in 1943 the light curves of Tycho's and Kepler's supernovae of 1572 and 1604 on the same graph. He was helped by the almost unbelievably accurate observations of the stars' brightnesses made by Brahe, Kepler and other seventeenth-century astronomers, with no instruments to help them apart from their eyes.

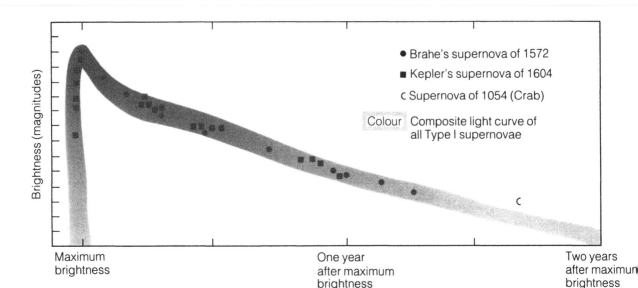

Brightness (magnitudes)

Maximum
brightness

One year
after maximum
brightness

Two years
after maximum
brightness

• Brahe's supernova of 1572
■ Kepler's supernova of 1604
c Supernova of 1054 (Crab)

Colour Composite light curve of
all Type I supernovae

FIG. 56. *Light curves of historical supernovae. The brightness of both Brahe's supernova of 1572 and Kepler's of 1604 follow the composite curve created from the light-curves of all Type I supernovae. The only discordant point is the point for the supernova of 1054, representing the brightness of the Crab supernova about a year and a half past maximum. This raises doubt that the Crab supernova was of Type I. However, few Type I's have been followed for more than a year past maximum, since those in external galaxies have faded into invisibility by then, and the curve is not well defined there.*

Light curve of Brahe's supernova

Astronomers describe the brightness of a star by a number they call its *magnitude*, but rather confusingly the magnitude scale seems to run the wrong way. Bright stars are said to be of the first magnitude, and the faintest ones which can be seen with the naked eye are sixth magnitude. Venus and Jupiter are brighter than first magnitude stars, and so actually have negative magnitudes. With a telescope one can perceive fainter and fainter stars which have bigger and bigger magnitudes. The faintest stars have magnitude 25. The brightest star seen was the supernova of 1006 which, at magnitude −8 to −10, was 100 million million times brighter.

Brahe's observations of the supernova of 1572 are drawn up in a form similar to that used by present-day astronomers to measure the brightness of variable stars. In so doing they cause starlight

to fall into a photometer, an instrument which produces an electric current indicating the star's brightness. By pointing the photometer at constant stars whose brightness or magnitude has already been determined and then at the variable star they see how the constant stars compare with the variable star. This is just what Tycho Brahe did, except that he used his eye, not an instrument, to compare the brightness. A feature of this method is that if at a later date more accurate magnitudes of the constant stars become available, the observations of the variable star can be reconverted into equally accurate magnitudes.

Thus, when Tycho said that in 1573 July and August the nova was equal to the principal stars in Cassiopeia, he himself deduced that it was third magnitude, this being the standard of brightness set down on the basis of those stars' appearance to Ptolemy and published in his book *Almagest*

(AD 144). The four stars in Cassiopeia which Ptolemy says were third magnitude are Alpha Cassiopeiae, whose magnitude measured with a photometer is 2.47, Beta Cassiopeiae at 2.42, Gamma Cassiopeiae at 2.25, and Delta Cassiopeiae at 2.80. The average is close to magnitude 2.5, so this was the magnitude of the supernova between 1573 July and August in the modern system of measuring magnitudes. The fact that the observations have been converted into modern stellar magnitudes 400 years after they were made, rather than the day after, is immaterial and a surprisingly accurate light curve of Brahe's supernova can be drawn up. He wrote:

When first seen the nova outshone all fixed stars, Vega and Sirius included. It was even a little brighter than Jupiter, then rising at sunset, so that it equaled Venus when this planet shines in its maximum brightness. The nova stayed at nearly this same brightness through almost the whole of November. On clear days it was seen by many observers in full daylight, even at noontime, a distinction otherwise reserved to Venus only. At night it often shone through clouds which blotted out all other stars.

However, the nova did not retain this extraordinary brightness throughout its whole apparition but faded gradually until it finally disappeared. The successive steps were as follows:

As already stated, the nova was as bright as Venus in November [1572]. In December it was about equal to Jupiter. In January [1573] it was a little fainter than Jupiter and surpassed considerably the brighter stars of the first class. In February and March it was as bright as the last-named group of stars. In April and May it was equal to the stars of the second magnitude. After a further decrease in June, it reached the third magnitude in July and August, when it was

closely equal to the brighter stars of Cassiopeia, which are assigned to this magnitude. Continuing its decrease in September, it became equal to the stars of the fourth magnitude in October and November. During the month of November, in particular, it was so similar in brightness to the nearby eleventh star of Cassiopeia that it was difficult to decide which of the two was the brighter. At the end of 1573 and in January 1574 the nova hardly exceeded the stars of the fifth magnitude. In February it reached the stars of the sixth and faintest class. Finally, in March, it became so faint that it could not be seen any more.

Walter Baade condensed this description to modern magnitudes and produced Table 3. Brahe then turned to Kepler's supernova of 1604.

Light curve of Kepler's supernova

The light curve of Kepler's supernova can also be well determined from the seventeenth-century observations because Mars, Jupiter and, later, Saturn, provided useful comparisons. At its maximum in October 1604, it was somewhat brighter than Jupiter, and it was still almost as bright as Jupiter when it disappeared near the Sun in November. In January, when it reappeared, it was only about as bright as Antares. It continued to become fainter until it reached the fifth magnitude in October 1605. By the following spring, it was no longer visible to the naked eye. Throughout its appearance Kepler made a series of comparisons between it and other stars, possibly modelled on Brahe's observations of the supernova of 1572.

Drawing the light curves of Tycho's and Kepler's supernovae on the same graph as the light curve of the supernova he observed in IC 4182, Baade found that all three objects are the same kind of supernova – Type I.

Table 3. *Magnitudes of Tycho's supernova*

Date	Days after maximum brightness	Description	Magnitude
1572			
Nov.	0	Almost as bright as Venus	−4.0
Dec.	30	About as bright as Jupiter	−2.4
1573			
Jan.	61	A little fainter than Jupiter	−1.4
Feb.–Mar.	107	Equal to brighter stars of first magnitude	+0.3
Apr.–May	167	Equal to second magnitude stars	+1.6
Jul.–Aug.	259	Equal to Alpha, Beta, Gamma and Delta Cassiopeiae	+2.5
Oct.–Nov.	351	Equal to stars of fourth magnitude	+4.0
Nov.	365	Equal to Kappa Cassopeiae	+4.2
1573			
Dec.–Jan.	412	Hardly brighter than fifth magnitude	+4.7
1574			
Feb.	457	Equal to stars of sixth magnitude	+5.3
Mar.	483	Became invisible	+6.0

The Crab – Type I or II?

About the supernova of 1054, the Crab Nebula supernova, there is, however, some doubt. The data are scanty but it seems that after SN (supernova) 1054's fairly quick decline to magnitude −3.5 it faded to magnitude +5 in 630 days, a drop of 8.5 magnitudes, compared with the norm of 10.5 magnitudes for the prototype Type I. Thus, writes Minkowski,

the difference of 2 mag between the supernova of 1054 and supernovae of Type I is not conclusive evidence in view of the many uncertainties, but it tends to contradict the interpretation of the supernova of 1054 as Type I and certainly does not make it mandatory.

This is important in trying to explain why the remnants of Tycho's and Kepler's supernovae are so different in their appearance from the Crab Nebula it would be easier to understand if the former were Type I and the latter Type II.

The true brightness of a supernova

The brightness of all Type I supernovae at their maximum is probably the same, insofar as this can be tested. There are three uncertainties which have to be considered. The first is that many supernovae are on the decline when first discovered and there is no way of determining what their maximum brightness was. Only a supernova whose maximum brightness is well established can be used in this test.

Second, supernovae are at varying distances from Earth so that their brightness as measured depends on how far away they are. Astronomers have to correct their measurement of the brightness of the supernova by calculating how much brighter it would appear if placed at some

Table 4.   *Magnitudes of Kepler's supernova*

Date	Days after maximum brightness	Description	Observers	Magnitude
1604				
Oct. 8	−9	Not seen	(several)	+3 or more
Oct. 9	−8	As bright as Mars	The physician	0.9
Oct. 10	−7	Like Mars	Capra Marius	0.5
Oct. 11	−6	Still brighter than Oct. 10	The physician	−0.7
Oct. 12	−5	Almost as bright as Jupiter	Roeslin	−1.5
Oct. 15	−2	As bright as Jupiter	The physician and Fabricius	−2.2
Oct. 17	0	Much brighter than Jupiter	Kepler	−2.6
1605				
Jan. 3	78	Brighter than Alpha Scorpii	Kepler	0.9
Jan. 13	88	Brighter than Alpha Bootis and Saturn	Kepler	0.0
Jan. 14	89	About as bright as Mars in Oct. 1604	Fabricius	0.9
Jan. 21	96	As bright as Alpha Scorpii	Maštlin	1.2
Jan.–Feb.	100	As bright as Alpha Virginis	Heydon	1.2
Mar. 20	156	Not much brighter than Zeta and Eta Ophiuchi	Kepler	2.4
Mar. 27	163	Same	Brengger	2.4
Mar. 28	164	Not much brighter than Eta Ophiuchi	Cristini	2.4
Apr. 12	179	As bright as Eta Ophiuchi	Fabricius	2.6
Apr. 21	188	Same	Kepler	2.6
Apr. 13	302	As bright as Xi Ophiuchi	Kepler	4.5
Aug. 29	318	About as bright as Xi Ophiuchi	Kepler	4.5
Sep. 13	333	Fainter than Xi Ophiuchi	Kepler	5.0
Oct. 8	356	Fainter than Xi Ophiuchi, difficult to see	Kepler	5.8

standard distance from Earth. (For various historical reasons this standard distance has been chosen as 10 parsecs, about 32.5 light years.) To make the calculation astronomers have to know at what distance the supernova lies. This can be done for the nearest galaxies by comparing the brightness of certain kinds of variable stars in those galaxies (cepheids) with cepheids in our Galaxy. Often, however, no individual stars can be distinguished in a galaxy in which a supernova occurs (save the supernova itself) and astronomers fall back on the so-called Hubble red-shift relation.

Edwin Hubble found in 1929 that on average all galaxies were receding from our Galaxy at speeds such that the more distant galaxies recede faster, and their spectra are red-shifted more, because of the Doppler effect. According to the best recent determinations 55 km/s is added on average to a galaxy's recessional speed for every million parsecs (3.25 million light years) distance it is from Earth. Thus, if a galaxy's speed is measured by the Doppler effect to be 5500 km/s, its distance is around 100 million parsecs. A supernova of Type I might have a maximum brightness of 16 in this galaxy but, when brought to a distance of just 10 parsecs (10 million times closer), it would be 10 million squared times brighter, that is 100 million million times brighter. Each factor of 100 in brightness corresponds to 5 magnitudes, so 100 million million means a difference of 35 magnitudes, and the supernova would appear at magnitude 16−35 = −19 if it were at a distance of 10 parsecs − far brighter than the Full Moon.

The third correction that ought to be made is for the amount of light from the supernova that is absorbed by dust in our Galaxy, and in the parent galaxy of the supernova. The absorption correction corresponding to our own Galaxy's dust can usually be made with adequate accuracy on the simple assumption that the dust forms a slab in our Galaxy, parallel to the Milky Way, through which astronomers point their telescopes towards supernovae in different directions, looking through different slanting thicknesses of dust. Astronomers have no information on the absorption in the parent galaxy, because they cannot tell whether the supernova is on the nearer or farther side of the galaxy.

The latest result of these calculations on the absolute magnitude of Type I supernovae is that the average magnitude is −19.5 to −20, corresponding to a luminosity nearly 10 billion times that of the Sun. This number is considerably in excess of the first estimates (in the 1930s) of the luminosity of supernovae because, at that time, the scale of the Universe was considerably

underestimated.

Knowing now that the absolute magnitude of Type I supernovae is about −19.5, we can attempt to calculate how far away such a supernova has to be pushed in order to appear magnitude −4.0 (Tycho's supernova) or −2.6 (Kepler's supernova). These calculations are somewhat handicapped by the correction for dust in our Galaxy. Since these two supernovae are within the Galaxy, the assumption that the absorption lies in a slab is no longer relevant (the supernovae might be within the slab). We can, instead, use the fact that dust in the Galaxy not only absorbs starlight, it reddens it too, just as the Sun is reddened at sunset when its sunlight passes slantwise through the Earth's atmosphere. If we can obtain information about the colour of the two early supernovae and know what colour (by present-day measurement of Type I supernovae in other galaxies) they might have appeared had there been no dust, then we can determine how much dust lies between Earth and the supernovae. We have to rely on subjective impressions by Tycho and Kepler of the colour of the respective supernovae and the argument treads on quicksand at this point.

However, all observers stressed that, at its maximum, Tycho's supernova had the yellow colour of Venus or Jupiter, becoming redder like Mars or Aldebaran after a month and whiter like Saturn at the same time thereafter. Kepler's supernova seems to have been even redder on the whole than Tycho's. On the basis of a discussion of these observations, Minkowski concluded that the absorption of light from Tycho's and Kepler's supernovae by dust in our Galaxy was 2.1 and 3.3 magnitudes respectively. Thus, had there been no dust, they would both have appeared at magnitude −6, or 14 magnitudes fainter than if at the standard distance of 32 light years. They are

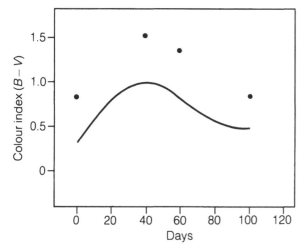

FIG. 57. *Colours of Tycho's supernova. In 1572 November, Tycho's supernova was the colour of Venus and Jupiter, in 1573 January like Mars and Aldebaran, in 1573 April like Saturn. Astronomers measure colour of stars and planets by a quantity called 'colour index', which is the difference between the magnitude B of the star seen through blue glass and the magnitude V seen through green-yellow glass. Blue stars have a small colour index, B–V, and redder ones a large colour index. A plot of the colour index of Tycho's supernova (as given by the colour index of the planets and stars to which it was compared) shows all the points to lie above the usual colour curve for supernovae of Type I. In other words, Tycho's supernova was redder than normal. This was reddening caused by the effects upon the light from the supernova by the dust through which the light had to travel.*

both, therefore, some 20 000 light years distant. We have to be wary of the large uncertainty of this estimate, caused mainly by the doubt about the reddening correction.

What about the light curves of the Type II supernovae? There is considerable diversity in these but after the maximum, Type IIs typically fade by 1.5 magnitudes and then almost halt their decline for 50 days. They then decline more rapidly at no particular rate, fading by two magnitudes in 60–120 days. Because Type IIs are more diverse, because they are rarer than Type Is, and because few have been well observed, their absolute magnitude is not known with certainty, but they are fainter than Type Is by a magnitude or two.

Supernovae as extragalactic probes

The brightness of supernovae makes them useful as probes of the intergalactic medium. As the light from a supernova passes from a distant galaxy to ours, it traverses the outer regions of the distant galaxy, intergalactic space and the interstellar space of our own Galaxy. Any material in the supernova's galaxy which leaves its traces in the supernova's spectrum can be identified because it will show the radial velocity of the distant galaxy. Material in the interstellar medium nearby to the Sun shows at a low radial velocity because our Sun and the galactic gas share the same general motion. Any material between shows at some intermediate velocity, and the exact value of the velocity at which the material is detected is the clue to where it is located along the line of sight. Thus the supernova reveals the presence of material in these intermediate regions, which cannot usually be investigated because stars in the normal range at these distances are too faint for such a delicate study.

The atoms and ions of gas in the far regions of our Galaxy and beyond are in their *ground state* – unexcited by any weak ambient radiation so far from any stars. Such atoms often betray their presence by absorbing ultraviolet radiation, and the ultraviolet spectra of extragalactic supernovae, recorded by the International Ultraviolet Explorer (IUE) satellite, show these effects. In the supernova of 1980 November in the spiral galaxy NGC6946, for example, which was a Type II supernova reaching magnitude 11.5, Max

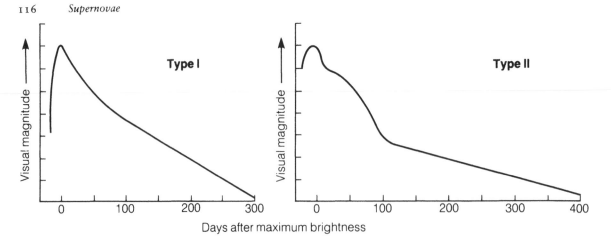

FIG. 58. *Light curves of Type I and Type II supernovae. The simpler light curve of the Type I shows a drop of just over 3 magnitudes from maximum brightness in 100 days, followed by a gentler decline totalling 7 magnitudes after 300 days. The more complex Type II light curve shows a faster drop of over 4 magnitudes in the first 100 days, followed by a much gentler decline totalling 7 magnitudes after 400 days. From data by Gustav Tamman.*

Pettini and a dozen collaborators from five countries detected absorption from magnesium, iron and manganese in the ultraviolet spectrum of the supernova as well as calcium and sodium in its optical spectrum.

The unexpected result was that interstellar gas exists in our Galaxy 30 000 light years above the galactic plane, undetected before these observations because (apart from stars in the Magellanic Clouds) no star beyond that distance was bright enough to study. Normally interstellar gas is thought of as being confined near to the plane of the Galaxy, within, say 1000 light years of the plane. The discovery has exciting implications for cosmology because it connects with the discovery of features in quasar spectra which appear to be due to big haloes of gas around other galaxies – galaxies are bigger than was thought! The supernova observations also showed absorption which could not be attributed to our Galaxy, NGC6946 or any other galaxy nearby. This absorption appears to come from unexplained intergalactic material.

Spot the difference

The original distinction between Type I and Type II supernovae was made by Minkowski by looking at their spectra. By 1941 he had taken spectra of 14 supernovae and found that 9 were very similar in showing various distinctive bands of colours, whereas spectra of the 5 examples of the second type were featureless during the period of maximum brightness and then developed distinctive features because of hydrogen. The Type IIs look most like common novae – their spectra have blue-shifted spectral lines, indicating that material is approaching Earth at speeds as high as 15 000 km/s at the time of the supernova outburst. Presumably these lines are caused by a shell of material ejected from the supernova, which of course on its near side will be approaching Earth. Type Is have less understandable spectra, but recently it has been established that they have the same basic cause, though the expansion speed of the shell of material reaches some 20 000 km/s.

The difference between Type I and Type II supernovae is not just the distinction between two engravings of the same postage stamp, a fascinating exercise with little meaning beyond itself. The clue to the significance of the difference is given by the kind of galaxies in which the two types of supernovae are found.

Not all galaxies are alike. Once Hubble had

FIG. 59. *Spiral galaxy.* NGC6744 *is a spiral galaxy seen face on. Its spiral arms wrap around its central region. The central nuclear bulge is like an elliptical galaxy, and supernovae of Type I occur in spirals and in ellipticals. Type II's occur only in spirals (and in the closely related irregular galaxies). Therefore Type II's are associated with spiral arm material. As can be seen in* NGC6744, *the arms of spiral galaxies contain patchy clumps of bright stars, and the implication is that Type II's are the explosions of such stars.* © *Anglo-Australian Telescope Board 1981.*

FIG. 60. *Elliptical galaxies. Such galaxies often occur in clusters (these are in the Fornax Cluster) and are rounded objects like footballs of various sorts. The upper galaxy is* NGC1399 *and is surrounded by a swarm of globular clusters like satellites. The lower is* NGC1404. *Supernovae of Type II never occur in galaxies like these, only Type I's. © Anglo-Australian Telescope Board 1981.*

established that galaxies were beyond the Milky Way (*extragalactic*) he went further. From the vast numbers of photographs which he accumulated, he established the existence of a series of galactic shapes, ranging from irregular conglomerations of dust, gas and bright young stars, through spiral galaxies with arms in which the dust, gas and bright young stars concentrate, to structureless spherical or elliptical balls of old, red stars with scarcely a trace of dust and gas. Although Hubble himself warned against the implication that this sequence of galaxian forms

represented the evolutionary life of a galaxy, it is now thought that there is some connection between the appearance of a galaxy and the age of the stars within it. Only in spiral galaxies and in irregular galaxies is there still a supply of dust and gas to form young stars now. For some reason star formation in elliptical and spherical galaxies has long ceased. The massive bright young stars in elliptical galaxies have all ceased to shine and only stars like the Sun still visibly survive in them.

When astronomers examine the frequency with which supernovae occur in various kinds of galaxies, they find that no Type II supernova has ever occurred in an elliptical galaxy, or even in the kinds of galaxies considered as intermediate in shape between the ellipticals and the spirals. Only in galaxies with clear spiral arms have Type IIs ever been seen. Type Is on the other hand appear in all kinds of galaxies but favour the elliptical galaxies. The implication is that Type II supernovae are the consequence of the evolution of the more massive stars (with masses, say, of 10 times the solar mass) whereas Type Is are caused by the evolution of stars typically of 1 solar mass (or less).

While most astronomers are happy with this notion about the precursors of Type IIs, many have expressed qualms about accepting the contrary implication for Type Is. Some have claimed that there must be a few very massive stars in elliptical galaxies, though not as many as in spirals, and that Type Is, like Type IIs, are both formed from massive stars. The origin of the more frequent type of supernova, the Type Is, remains obscure. The most favoured theory is, however, that they come from exploding white dwarfs in close double stars.

9

The making of a neutron star

The stars, like measles, fade at last.

SAMUEL HOFFENSTEIN

So far we've seen how astronomers have demonstrated observationally that massive stars produce supernovae and that supernovae make neutron stars. Knowing that this occurs is not enough; astronomers want to know why. To understand why we have to look at the life history of stars in some detail.

The life and death of stars

The modern theory of stellar evolution goes like this. All over the Galaxy there are clouds of interstellar hydrogen: vague, tenuous, gaseous masses, called *nebulae* when they shine (which they do if they are near a bright star). For reasons which are not entirely clear, especially dense parts form within a nebula, and these dense parts have a sufficiently strong mutual gravitational attraction to make them fall together in larger and larger lumps called *protostars*. They are 'about-to-be' stars. Gravity is the force which attracts these lumps to one another, just as the force of gravity on Earth pulls all objects down to the surface, drawing them inexorably towards the centre.

As the gas of which the protostar is made is packed closer and closer in towards its centre, it begins to heat up, just as the air in a bicycle pump heats as it is compressed into the tyre. You can feel the valve getting hot as you pump. Such heating causes the star to stop collapsing. It causes the pressure in the centre of the star to increase, and this pressure shoulders the atoms of gas apart which, in turn, keep the outer layers from collapsing. The star becomes a finely balanced mechanism, tending to collapse into itself because of its own gravitational force, but prevented from doing so by the pressure of the gas inside.

This balance comes about only when the gas in the centre of the star reaches a temperature of many millions of degrees. In a star like the Sun, the centre reaches a temperature of 15 000 000 °C and a density of about 160 times that of water. In the innermost 3% of the volume of a star like the Sun, into which is packed two-thirds of the mass of the star, the temperature and the density are so high that nuclear reactions take place. Over the whole star, the hydrogen atoms from which the star is made have been split apart into their component bits by the force of the collisions between the atoms as a result of the high temperature; the hydrogen atom is so simple that it produces only two pieces, an electron and a proton, the massive central nucleus. In the very centre of the star the protons themselves have been forced so close together that there is a substantial chance that four protons will be able to stick together in a new arrangement. With the addition of two electrons this forms an alpha particle which is the central massive part, or nucleus, of a helium atom. As the four hydrogen nuclei are converted into a helium nucleus, energy is released, mostly in the form of gamma rays, which are very energetic light particles or photons, similar to light but with a million times more energy.

The gamma rays slowly diffuse up out of the star, being degraded on the way to a much larger number of lower energy photons which ultimately leave the surface of the star mostly in the form of light and infrared radiation. It is this light which leaves the star, travels through space, and permits stars to be seen. Indeed, today's sunlight is the result of gamma rays emitted from the centre of the Sun millions of years ago, released by the nuclear reactions, and providing a fossil record of them.

The energy production within a star can be thought of as brilliantly shining proof of Einstein's famous equation $E = mc^2$ – that is, mass and energy are equivalent, and one can be converted into the other given the right conditions. Each time four protons are converted into a helium nucleus, 0.7% of the mass of the four individual protons is converted into radiative energy. Although each nuclear event converts just a small amount of mass to energy, in total a star like the Sun radiates 4 million tons of its mass away during every second of its life.

At this rate of consumption even a mass as large as that of the Sun must be appreciably diminished in the course of time and less hydrogen is available to sustain the rate of energy production. The star becomes unbalanced, unable to support itself against its own gravitational attraction. The centre of the star begins to contract and the outer parts begin to expand, producing a *red giant star*.

In a Sun-like star, this occurs at an age of 10 billion years. The conversion of the four hydrogen protons to a helium nucleus still takes place, not in the centre but in a shell around the centre of the red giant. The centre of the star, which is by now largely helium, releases energy first by contracting but, as it contracts (if it is large enough) it eventually becomes hot and dense enough for a further kind of nuclear reaction to begin. The star begins to convert the helium nuclei into carbon nuclei by what is known as the *triple-alpha process*, because it involves the uniting of three helium nuclei or alpha particles to make a carbon nucleus.

This further conversion of energy in the triple-alpha process halts the contraction of the centre of the star for a while, but not for long. Because the triple-alpha process is a much less efficient means of generating energy than the conversion of hydrogen to helium, the star remains a red giant for only a comparatively short time. The star's central regions contract even more and, if the star is massive enough, the contraction causes the centre to heat up enough for yet further nuclear reactions to occur. In these, helium nuclei are successively added to carbon nuclei to form heavier nuclei like oxygen, neon, magnesium and silicon, possibly up to iron nuclei. Less and less energy is available from these reactions; they only briefly postpone the ultimate collapse of the centre of the star to a very dense state, known as *degenerate matter*, when the red giant star turns into a *white dwarf*.

Because it has collapsed so much, the density of a white dwarf is very high: about 1 000 000 g/cm³ so that a block 1 cm on each side – roughly the size of a sugar lump – weighs 1 ton. Left to itself, the white dwarf gradually cools off, rapidly at first from white hot to yellow hot, but then more slowly to red hot. As it becomes redder, it fades until, ultimately, it disappears from view. At the last stage, it is known as a *black dwarf*.

In 1932 the Indian-born American astrophysicist S. Chandrasekhar proved a remarkable theorem about white dwarfs, showing that no white dwarf could exceed a mass of about 1.5 times the mass of the Sun. A star of larger mass which attempted to become a white dwarf would not be able to hold itself up against the force of gravity and would have to turn into something other than a white dwarf. This maximum mass for a white dwarf is known as the *Chandrasekhar Limit*. The Chandrasekhar Limit lies between 1.44 and 1.76 solar masses, depending on precisely which nuclei the star is made of. If it is composed of helium nuclei, the maximum possible mass is 1.44 solar masses, and if of iron nuclei, 1.76 solar masses. Yet the number of white dwarfs known is too large for them to have evolved only from stars which are less than the Chandrasekhar limiting mass. (Near the Sun there are about five white dwarfs in every cube of space whose side is 30 light years.) There is even one white dwarf in the Pleiades star cluster, a cluster so young that only stars 6 times the mass of the Sun can have had time to evolve to white dwarfs. Somehow, stars which are heavier than the Chandrasekhar limiting mass must lose material before they can become white dwarfs.

Mass loss

Stars can lose material at late stages in their evolution, while they are red giants. Because a red giant is very extended, its mass is spread over a large volume and it has a low surface gravity. Bits of the atmosphere can, relatively easily, be thrown back into space by various storms, flares and winds from the star's surface. The winds were first studied in red giants by Armin Deutsch in 1956, but the method available to him was not fine enough to show that *all* red giants lost mass. That they do became clear with the development of microwave and infrared techniques. Ed Ney and Neville Woolf detected infrared emission from the graphite dust grains which condensed out of the atoms of carbon in the outflowing gaseous atmospheres of red giants. Microwave emission was found from hydroxyl residues, water and carbon monoxide molecules (simple chemical structures made from a few atoms of the most common elements – hydrogen, oxygen and carbon – in a red giant's atmosphere). It was even possible, because these long wavelength emissions penetrated the dust, to discover by these techniques red giant stars whose visible light was hidden by circumstellar shrouds of the stars' own making.

These techniques showed that the circumstellar haloes around red giants are very massive. Andrew Bernat and his colleagues at Kitt Peak have shown that Betelgeuse has a halo 3 arc minutes in diameter, and the star CW Leonis has one twice this size. In real terms this halo is 1000 times the diameter of our own solar system, and contains at least two solar masses of material. It seems that stars of up to 8 solar masses can lose enough matter by stellar winds to drop below the critical Chandrasekhar Limit of 1.4 solar masses and therefore become white dwarfs. In fact the stage between red giant and white dwarf is marked by the dramatically beautiful production by the star of a *planetary nebula*. The name arises because, when seen through a small telescope, the planetary nebulae have the appearance of a flat disc like the planets Uranus and Neptune. The nebulae have at their centres hot stars, with the appearance of being hot, bright, white dwarfs. A typical planetary nebula is comparable in size to the solar system, but may be much larger; many bright planetary nebulae are surrounded by fainter haloes. The Canadian astrophysicist Sun Kwok visualizes the planetary nebula as part of the halo of stellar wind material rendered visible by the light from the bright core of the red giant as it evolves into a white dwarf.

Through these processes, stars of mass 4 times the Sun's mass, and perhaps up to about 8 solar masses, can become white dwarfs. Stars larger than this cannot cope with the problem of losing enough mass to turn into white dwarfs. It appears that these are the stars which turn into Type II supernovae. Roger Blandford of the University of California and his colleagues believe there is evidence for this from the Crab Nebula's precursor star. Running into the north of the Crab is a 'jet', which they interpret as a hollow cylinder of material created by the star as it

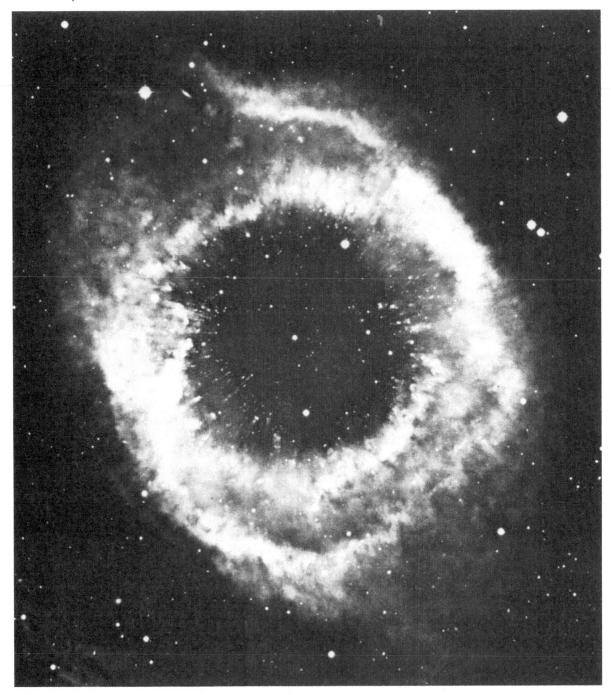

FIG. 61. *Helix Nebula. The nearest planetary nebula to the Sun, the Helix Nebula is centred on a blue star. Faint material surrounds the bright complete circular structure which gives planetary nebulae their name. The blue star was originally more massive than the Sun, but lost its outer regions in a stellar wind. It is now about the same mass as the Sun, and on its way to become a white dwarf. The blue colour of the star represents its temperature and it is now so hot that it has caused its lost circumstellar envelope to shine in this dramatic way. Photograph by David Malin © Anglo-Australian Telescope Board.*

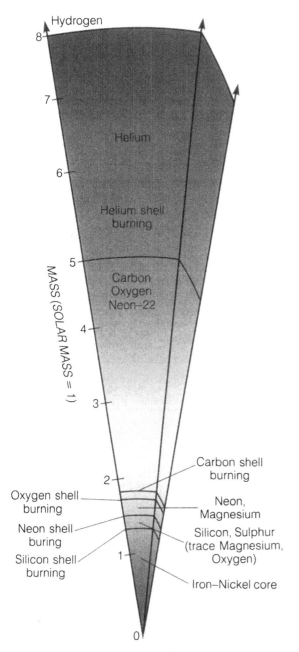

Hydrogen

Helium

Helium shell burning

Carbon Oxygen Neon–22

MASS (SOLAR MASS = 1)

Carbon shell burning

Oxygen shell burning

Neon, Magnesium

Neon shell buring

Silicon, Sulphur (trace Magnesium, Oxygen)

Silicon shell burning

Iron–Nickel core

FIG. 62. *Structure of a 20 solar mass Type II supernova precursor. The star is a series of onion skins of different material which burn in layers at the junctions of the skins. The structure of the star is shown in this radial section, in which the radial dimension has been distorted to represent uniformly the mass of the star enclosed at that radius. This plot emphasizes the interior portions of the star, which in true radial distance scale are much smaller than this plot implies. Only the tiny central iron-nickel core survives the explosion.*

moved through space. They envisage the material flowing out from the star and being halted in the interstellar material. They estimate the visible material at 6% of the mass of the Sun and that the jet was produced over the last 300 000 years.

Core collapse

The Crab's precursor did not lose enough mass to get below the critical mass, so its core could not evolve to a white dwarf. Exactly what happened is not clear but it certainly was a catastrophe. The calculated scenario for the supernova explosion of a 20 solar mass star is as follows.

Because of the successive burning of heavier and heavier nuclear fuels, the interior of the star, just before it goes supernova, is made of a series of concentric shells of different nuclei. Its structure is like that of an onion, the inner shells containing the heavier nuclei. At the centre of the star is an iron–nickel core of 1.5 solar masses. Surrounding it is a small zone of silicon and sulphur. Outside this mantle lies another small zone of neon and magnesium. 3 solar masses of carbon, oxygen and neon follow, then 3 solar masses of helium, and the remaining 12 solar masses on the outside of the star are of hydrogen. At the boundaries between each zone nuclear burning is taking place, but the fuel eventually gives out and the inner core collapses as gravity finally wins over the dying pressure inside the star. The collapsing core shrinks so that its density increases beyond even the degenerate density of a white dwarf. It becomes, in fact, a *neutron star*, since electrons in the iron–nickel core are forced into the protons present in the iron and nickel nuclei. The negative electric charge of the electrons cancels out the positive charge of the protons, giving electrically neutral particles called *neutrons*.

Neutron stars are subject to a limit on their mass similar to the Chandrasekhar Limit on the mass of a white dwarf. There is some discussion about the size of the critical mass of a neutron star, but C. E. Rhoades and R. Ruffini, have calculated that the critical mass certainly cannot exceed 3.2 times the mass of the Sun. All neutron stars have masses below this figure, probably below 2.5 solar masses. A typical neutron star has a mass 1.5 times that of the Sun.

The density of neutron stars is extremely great, at least 10 million million grams per cubic centimetre, so that a piece the size of a sugar cube would weigh 10 million tons, comparable with all the materials used to build a city the size of London or New York. This is because material to the amount of about 1.5 solar masses is packed into a star whose diameter may be only 20 kilometres.

The supernova explosion

If the end product of a 20 solar mass star is a 1.5 solar mass neutron star, and something like 5 solar masses have been ejected from the precursor by mass-loss in the red giant stage, then it follows that about 10 solar masses of the outer parts of the star have yet to be accounted for. This is the material of the supernova explosion.

The outflow of this material at 15 000 km/s or so causes the surface area of the 'star' to grow rapidly. The light from a star is brighter when its temperature is higher and its surface area is larger, and the rapid growth of surface area is the cause of the initial brightening which we perceive as the supernova explosion. At such a high speed of outflow, the star grows to the size of the solar system (the diameter of Jupiter's orbit) in 1 day, and its surface area becomes more than a million times that of the Sun. For 1 week or so this growth of surface area continues to cause the supernova to brighten but, at the same time, the star is cooling because of its expansion. For a while, the growth of surface area and cooling of the surface compensate, and the supernova

remains at maximum brightness. Eventually cooling becomes the dominant effect, and the supernova begins its decline.

The question which is as yet still unanswered is this: how does the gravitational collapse of the core get turned around into the explosion whose results we so dramatically see? The outer parts of the star begin to free-fall when the core collapses to a neutron star. The whole star's support is removed virtually instantaneously, since the inner core collapses in about 1 thousandth of a second. What mechanism can there be to reverse the fall? What does the collapsing core generate which pushes out the in-falling layers? Between 1966 and 1979 astronomers believed they had found the answer in neutrinos.

Neutrinos

Most things in science which we know about have been *discovered* by somebody. Yet it is possible to talk about some things actually being *invented*, in the sense that a theoretical scientist saw that logically they had to exist before they had been discovered. This is so in the case of the neutrino, which was invented by Wolfgang Pauli in 1930. The story of the neutrino is also a good example of the way in which the behaviour of submicroscopic particles, individually all but undetectable, can have devastating effects on a large scale. For, apparently, neutrinos can actually cause a supernova.

Pauli had to invent the neutrino because of a fundamental law of science: energy can be changed from one form to another, but it never appears or disappears. Atomic scientists in the 1920s were worried that this law was apparently being broken in a certain kind of nuclear transformation called beta-decay; this is, essentially, the way in which a neutron spontaneously decomposes to form a proton and an electron. A proton is positively charged and an electron is negatively charged, while a neutron has no charge at all. The result of a neutron suffering beta-decay is still no net electric charge, since the charges on the electron and the proton cancel each other. Therefore, electric charge did not appear or disappear in beta-decay: in the jargon of nuclear scientists, it was conserved. Although the charge was conserved, the problem was that the total energy of the neutron alone was greater than that of the pieces after beta-decay. Energy was not being conserved: it seemed to disappear.

To get over this difficulty, Pauli invented a particle which could have no charge, but which would balance out the energy equation. A small fraction of the energy in beta-decay, he said, was carried off by this imagined particle which no one had ever detected. Because the energy carried away was sometimes very small, the neutrino had to have a very small mass – even zero! Pauli was aware that he was dangerously near sophistry. He thought that no one would ever detect the neutrino and told the astronomer Walter Baade:

Today I have done the worst thing for a theoretical physicist. I have invented something which can never be detected experimentally.

Pauli originally called his imaginary particles *neutrons*, but they were different from what we now call neutrons, which were not actually discovered until 1932. Enrico Fermi, with exasperation and gestures to match, explained the difference to an audience of slow physicists at a conference in 1933:

The neutrons discovered by Chadwick are big, Pauli's neutrons were small. They should be called neutrinos.

The *-ino* ending in Italian is a diminutive, like *bambino*, and the name stuck.

Fermi worked out that the chances of a given neutrino reacting with anything were very small. If a neutrino travels at the speed of light through a 3000-light-year-thick slab of matter with the density of water (the average density of the Sun), it has only a 50–50 chance of reacting with a proton.

Nonetheless, neutrinos are given off in large numbers from nuclear reactors: every second 10 trillion pass through a square centimetre near the reactor. Although each one has only a small chance of being detected in experiments, there are so many available that just a few actually are. Pauli was too pessimistic – the existence of neutrinos has been confirmed.

Stars like the Sun are vast nuclear reactors: the Sun creates in its centre vast numbers of neutrinos every second. Because the Sun's radius is only 2 light seconds, very small compared with 3000 light years, nearly all the neutrinos created by the Sun dash unheedingly out of the Sun's surface and diffuse through space. By the time they reach Earth, the number passing through each square centimetre per second has been calculated to be 65 billion: a much smaller number than can be made in a terrestrial nuclear reactor, but still just enough to give a few detectable interactions.

Neutrinos from supernovae?

The Sun is a weak source of neutrinos compared with a supernova. Even before it forms a neutron star, the precursor of a supernova creates neutrinos in abundance by two main methods. In the first method, radiation creates matter. It does so in a beautifully symmetric way. To every kind of particle of matter there corresponds a kind of particle of antimatter and, if it is energetic enough, radiation can produce matter–antimatter pairs of particles. A gamma ray produced at the centre of a massive star is energetic enough to create an electron and its antiparticle, a positron. These can recombine and produce a pair of neutrinos. In the second method, an electron is captured by a proton in a nucleus with the emission of a neutrino, and the resulting neutron decays back to an electron, a proton and a neutrino. The neutrinos run off with a fraction of the energy, but the original nucleus still remains to suffer again this process of attrition. This is called the *Urca process*, after a casino in Rio de Janeiro, where the customer loses little by little.

The energy carried away by neutrinos from the centre of a massive star is the very cause of the supernova explosion itself. Energy transformed into speeding neutrinos is lost from the star virtually instantaneously, whereas energy transformed into radiation jostles its way out of the star and helps to support the star against its own gravitational pull. The more energy lost from the centre of the star as neutrinos, the less support the released energy gives to the star. When support drops too far, collapse becomes inevitable.

Initially, about half of the particles in the centre of the star are neutrons; the other half are protons. Both neutrons and protons swim in a sea of electrons. The implosion forces the protons to swallow the electrons and make neutrons: the star centre becomes a neutron star. Each creation of a neutron liberates one neutrino, further increasing the neutrino output.

In 1974, as the significance of the flood of neutrinos from supernovae began to be appreciated. Astronomers realized that, in spite of the fact that neutrinos were so reluctant to interact with anything, there were so many released from the collapsing core that the tiny fraction which did get caught in the in-falling outer layers had a significant effect on the fall;

perhaps they deposited sufficient energy into the in-falling layers to reverse the fall into an explosion. Perhaps, paradoxically, astronomers should look for the cause of the most powerful explosions known in the particle most unlikely to push anything at all.

At first the calculations were encouraging. Apparently as the core itself collapsed, neutrino production rose so rapidly that the resulting pressure caused the core to rebound. The bounce of the core at its edge some 10–20 km from the star's centre generated a powerful shock wave which travelled outwards, met the in-falling outer layers and reversed the fall into an explosion. Later calculations in 1979 by Hans Bethe and his collaborators (known as the BABBLY group: H. Bethe, J. Applegate, G. Baym, G. Brown, J. Lattimer and A. Yahil) did not confirm the details of this idea, but still seemed to show that the core bounced and that a shock wave travelled outwards. But it was also found that, as the shock wave travelled into regions of lower density, insufficient numbers of the neutrinos were picked up by the in-falling matter. In fact, the neutrinos escaped from the star in such numbers that a rarefied zone of low pressure occurred behind the shock and sucked back the rebounding shock wave, creating in-fall again.

Astrophysicists so desperately need to reverse this fall that, in 1981, David Schramm and John Wilson tried to use another invented particle to dump energy into the in-fall. The *axion* was postulated to be a particle created in the core, with the ability to leave it (like neutrinos), but with a significant difference from neutrinos – its lifetime was short. It decayed into photons as it reached the outer layers of the core, released its energy in the in-falling matter and created an explosion in the core-bounce shock wave. Schramm and Wilson pointed out that these

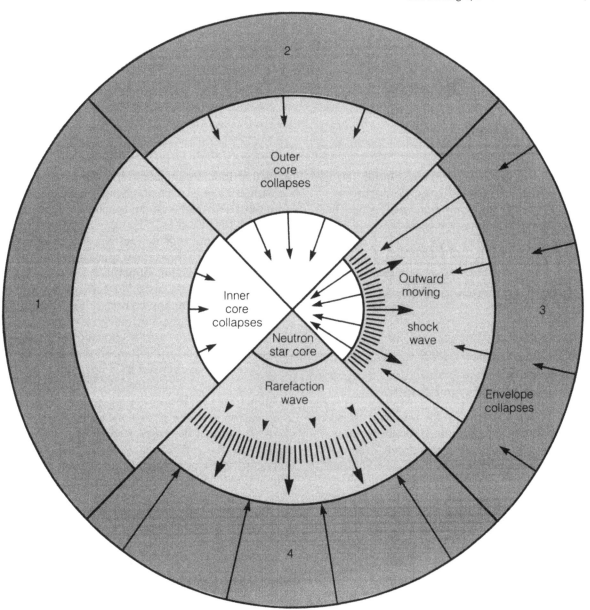

FIG. 63. *Core collapse. The cross-section of a star is shown at four glimpses of its collapse. 1. In sector 1, the inner core has exhausted its energy and cannot sustain itself by internal pressure against its own gravity. It collapses. 2. This removes support from the outer core, which also collapses. The inner core is in the process of being converted into neutrons. 3. The pressure of neutrinos released from the neutron core causes the core to bounce. A shockwave moves outwards into the in-falling outer core. Meanwhile the envelope of the star begins to collapse. 4. As the neutron star core is formed, a rarefaction wave forms behind the shock. Somehow the processes in this region reverse the infall of the envelope and outer core. They are ejected into space as a supernova explosion.*

particles 'would cause a dramatic solution to the long-standing gravitational collapse supernova problem', even though they were well aware that axions were almost totally speculative.

Supernovae deserve to be understood,

says Bob Kirshner, a University of Michigan astronomer, in his *Optimist's Guide to Supernovae.*

While the usual cheerful anarchy still prevails, something must be right with the general picture.

How spinning stars speed up

A curious thing happens to the rotational period of a star which collapses into a smaller volume. Before collapse, the star rotates slowly – all stars rotate to a greater or lesser extent. The Sun, for example, spins once on its axis every 25 days 9 hours, just as the Earth spins on its axis in one day. Stars more massive than the Sun generally rotate faster, and periods as short as half a day are known for some stars. When stars expand and become red giants, their period of rotation lengthens: they slow down. During their subsequent collapse to white dwarfs or neutron stars, they speed up again.

The reason for this is a physical law called the *conservation of angular momentum*, which says that in this kind of situation the cross sectional area of a star divided by its rotational period will remain roughly a constant. When the star expands in the red giant stage its cross sectional area gets larger and therefore its period will become proportionately greater – the star slows down. When the star subsequently shrinks, its cross sectional area becomes smaller, its period is reduced in proportion – and the star speeds up.

Ice skaters exploit the law of the conservation of angular momentum when they pirouette on the ice with arms outstretched, giving themselves a large cross sectional area, and then rapidly bring their arms to their sides. This decreases their cross sectional area and thus decreases their period of rotation; in other words, they spin faster.

Consider a star like the Sun with a cross sectional area of 600 billion square miles and a period of about 1 month (2 million seconds). If it were possible to collapse the Sun suddenly to the size of a neutron star, the cross sectional area would decrease by a factor of a billion so that the period too would have to decrease by a factor of a billion, to about a 5 hundredth of a second. In a real supernova explosion not all the star collapses to a neutron star; the parts which are ejected carry off some of the angular speed of rotation so that, in practice, neutron stars rotate more slowly than this – their periods are several times longer although still almost incredibly fast. The Crab Nebula pulsar, when it started to rotate, must have been spinning about twice as fast as it is now, with a period of about a sixtieth of a second.

The rotation of heavy objects makes a very good repetitive clock. After all, the Earth was used as a clock until very recently when atomic clocks revealed the small irregularities in its rotation. The regular rhythms of a pulsar, and the Crab Nebula pulsar's rate of repetition of a few tenths of a second, led Thomas Gold to suggest that pulsars were rotating neutron stars. In some way, he suggested, a pulsar emitted a beam of light like a lighthouse; this beam, as it turned towards the Earth with each rotation, produced the pulses observed by radio and optical astronomers. When he found that the Crab pulsar was slowing down, feeding its pulses from its rotational energy, he became convinced that his idea was correct.

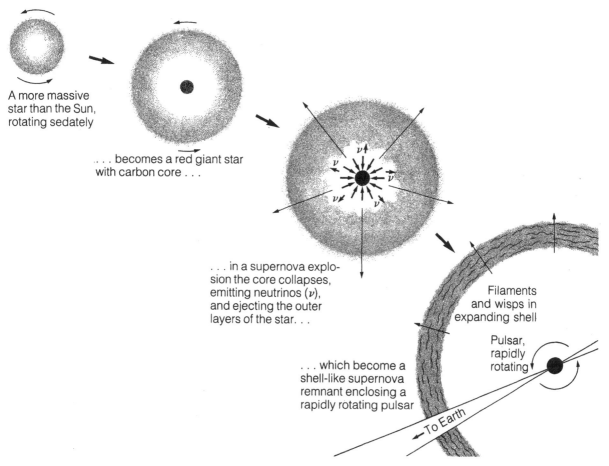

A more massive
star than the Sun,
rotating sedately

. . . becomes a red giant star
with carbon core . . .

ν
ν
ν
ν
ν

. . . in a supernova explosion the core collapses,
emitting neutrinos (ν),
and ejecting the outer
layers of the star. . .

Filaments
and wisps in
expanding shell

Pulsar,
rapidly
rotating

. . . which become a
shell-like supernova
remnant enclosing a
rapidly rotating pulsar

To Earth

FIG. 64. *Formation of a pulsar. The slow rotation of a normal star is converted
into the rapid rotation of a small neutron star.*

How pulsars may shine

It is not quite enough to say that pulsars are like lighthouses, beaming their light and radio rays into space as they spin around. How do they actually make the beam? The answer is unclear but tied up with an immense magnetic field at the surface of a pulsar. All stars have magnetic fields to a greater or lesser degree; when a star produces a pulsar its magnetic field is caught fast and throttled, squeezed into the smaller cross section of the neutron star. When this happens, the strength of the magnetic field increases enormously, by a factor of millions.

Pulsar astronomers assume that the magnetic field of the pulsar does not lie exactly along the polar axis of the pulsar. As in the Earth, whose magnetic axis leaves its surface in Canada, 15° of latitude from the North Pole, the magnetic field of a pulsar may lie off the pole of rotation, possibly even at the pulsar's equator. The pulsar is therefore a powerful electric generator, spinning a magnetic field lying across a rotor, and an electric current pours from the surface of the neutron star. Caught in the magnetic field, the electrons which make up the current stream out of the magnetic poles, radiating as they do so. The beam of radio waves which they emit points out of the magnetic axis of the pulsar and as the magnetic axis sweeps in the direction of the Earth we perceive a pulse. Possibly, we are in a position to see a diminished amount of radiation from the magnetic pole on the far side of the pulsar and then we can see two pulses; many pulsars, including that of the Crab, show two pulses during every rotation.

Perhaps not all pulsars shine in this way. Perhaps, in some, the beam of radiation is produced by electrons bunched not at the magnetic poles but off the surface of the neutron star near to the place where the magnetic field is moving at speeds close to the speed of light. Then the radiation is caused not by electrons beaming along the magnetic field but along the direction in which they are moving, at right angles to the magnetic field.

Whatever the cause, investigation of the way pulsars shine has stimulated research into the fundamental physics of fast-moving strong magnetic fields and particles. Only in the cosmic laboratory has it been possible to see such an experiment in operation.

Radio supernovae

Radio astronomers' studies of supernovae have traditionally concentrated on supernova remnants, but in the past few years three extragalactic Type II and one rather peculiar Type I supernovae have been detected by radio astronomers simultaneously with the optical outbursts. They were SN 1970g in the galaxy M101, SN 1979c in M100, SN 1980k in NGC6946 and SN 1983n in M83. The radio supernovae were all 10–100 times brighter than the radio supernova remnants in these galaxies. Like the optical supernovae, they flared up for just a few months and then faded away, presumably to form true radio supernova remnants in the future. In fact, retrospective searches for radio emission at the positions of supernovae which have appeared in the past have always been failures, so that even if some supernovae are very bright in the radio immediately after outburst they must fade away quickly. If a radio supernova occurs in our Galaxy then, for a while, it will be the brightest radio object in the radio sky, exceeding Cas A by a factor of 100 or so. This suggests straight away that, in the era of radio astronomy (say, since World War II), there has not been a radio supernova in our Galaxy, or it would have been noticed.

How the radio outburst is produced is not clear. One idea is that the radio emission is from a central pulsar, supplying relativistic electrons and a magnetic field to the gas created by the supernova. It is not understood how the radio waves propagate out of the plasma – they should be trapped inside, just as the ionosphere of the Earth traps short radio waves – but attributing the emission to a pulsar fits in with the idea that Type II supernovae make neutron stars and only Type IIs (and one rather peculiar Type I) have been detected as radio supernovae.

Starquakes

Although pulsars are very good clocks, rotating very regularly (apart from a gradual slowdown), most pulsars – particularly the two fastest, the Crab Nebula pulsar and the pulsar in the Vela supernova remnant – do, when examined closely enough, show irregularities. At the end of September 1969, for example, the period of the Crab pulsar suddenly decreased by one-third of a billionth of a second. There have been much larger jumps in the Vela pulsar's period, each of 200 billionths of a second. The reason for these sudden so-called glitches seems to be that neutron stars have a crust, just as the Earth does. Because they are rotating so fast, the crust is not precisely spherical but is flat at its poles and bulges at the equator, due to the centrifugal force there, just as the Earth is tangerine-shaped. As the pulsar slows down the centrifugal force becomes less, but the crust takes the strain for a time even though the equatorial bulge of the pulsar tries to fall. The crust cannot survive the stress indefinitely. It breaks and suddenly drops. By the law of the conservation of angular momentum previously mentioned, the pulsar suddenly spins faster.

The amount by which the thickness of the crust drops is minute – it is measured in millimetres! Yet this is enough to have a noticeable effect on the pulsar's period, just because it is usually so regular.

Astronomers call the sudden changes in the neutron star *starquakes*. The energy released in a starquake is enormous and flows out from the star into the space surrounding it. After the 1969 starquake on the Crab pulsar, a wave of energy was seen to flow outwards, rippling through the centre of the Crab Nebula. This, apparently, is the cause of the waves of activity which Baade noted and was one of the clues which convinced him correctly in 1945 that he had identified the Crab Nebula supernova's stellar remains.

10

Supernovae in binary stars

Only about 15% of all stars are in what we might call the 'lone star state' – that is, they are single (like the Sun). Nearly half (46%) have partners, with one star orbiting the other; the remainder (39%) occur in multiple star systems. It must, therefore, be common for a supernova to occur in a multiple star system.

Runaway stars

The star which explodes first in a double or binary star system is the larger of the two since more massive stars expend their nuclear energy at a faster rate than do smaller ones, and pass more rapidly to the late stages of stellar evolution.

What happens when the more massive star of a pair explodes as a supernova? Certainly the supernova explosion has some effect. In the first place, the smaller companion star receives the impact of the explosion and recoils like a wicket-keeper's glove or a catcher's mitt grasping a fast-thrown ball. If the companion star was describing a circular orbit about the exploding star, the orbit becomes eccentric or flattened to some degree. The companion may also be blasted by stellar shrapnel; material from the exploding star may significantly increase the mass of its companion. Even if it had been heading to a quiet end as a white dwarf before the supernova explosion, the companion may now itself become too large for this and thus be destined to become a supernova too. In this sense, at least, supernovae can cause supernovae.

Not all the exploding material accretes on to the companion star. Perhaps most is ejected into space. It is even possible that the blast from a supernova is so severe that it strips the outer layers from its companion, decreasing its mass and lengthening its life. All in all, more than half the total mass of the double star system can be blown off into space. In this case the pair is split

asunder, each star flying away like a stone flung from a slingshot. As a numerical example, consider a 30 solar mass star and a 4 solar mass star orbiting each other, making a total of 34 solar masses. The large star explodes leaving a 2 solar mass neutron star behind and ejecting 28 solar masses. 8 solar masses of this may fall onto the smaller companion, now making it 12 solar masses, while 20 solar masses of material is ejected into space. Because this is more than half the original mass of the double star, the neutron star and the 12 solar mass star are flung in opposite directions, speeding away from each other. Where there was once a double star, now there is simply an empty space at the centre of an expanding shell of fragments, with two stars rushing away from each other.

Where might we find such events? Stars of 30 solar masses are profligate with their nuclear energy and have lifetimes measured in only a few millions of years, compared with the 10 billion year lifetime of the Galaxy. They are thus young objects and must have been recently formed from the interstellar gas in the Galaxy. This gas only occurs in the plane of the Galaxy in spiral arms. Thus bright, massive stars are found near the galactic plane.

If one star of a pair goes supernova and disrupts the double star, the two stars are quite likely to be flung out of the galactic plane altogether. We would expect to see some pulsars moving at large speeds away from the galactic plane. The largest speed of the 26 pulsars whose proper motions have been measured in a clever Jodrell Bank survey was 400 km/s. Andrew Lyne selected for measurement some pulsars which happened by chance to lie near in the line of sight to a quasar, and measured on several occasions the position of the pulsar relative to the quasar in order to plot the motion of the pulsar across the

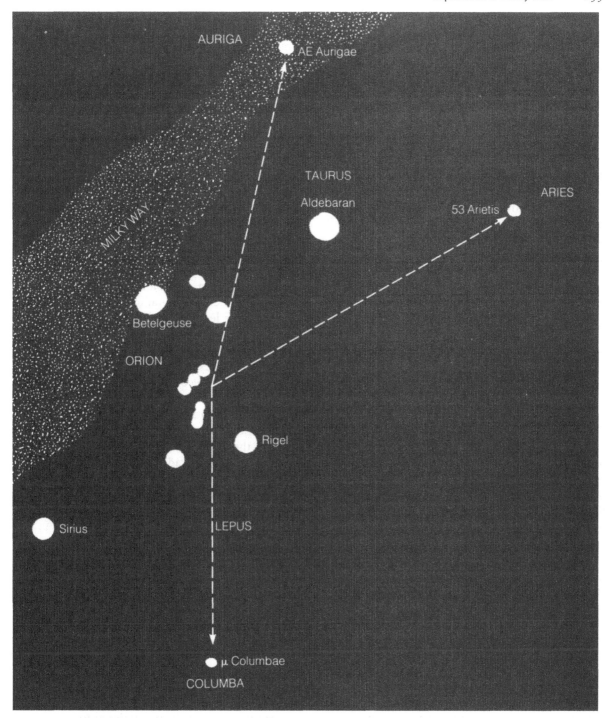

FIG. 65. *Runaway stars. AE Aurigae, 53 Arietis and Mu Columbae were thrown like slingshots from the constellation Orion three million years ago. Amongst the many massive stars in Orion one must have been part of a quadruple system. When it exploded, the other three stars of the system were ejected from the region. Each has travelled right across an intervening constellation to reach the present day positions 30 and 40 degrees from their origin.*

sky. His results, which show that pulsars generally are leaving the galactic plane at high speeds, are consistent with the idea that a supernova explosion which makes a pulsar in a binary star usually disrupts the binary star.

What of the former companions of the pulsars? Among the normal slow-moving population of bright, massive stars in our Galaxy there are some so-called *runaway stars*, with high speeds often exceeding 100 km/s. Three of these are Mu Columbae, AE Aurigae and 53 Arietis. From their speeds and directions, astronomers calculate that each left the region of the constellation of Orion some 3 million years ago, though they are now in the widely separated constellations of Columba, Auriga and Aries. Orion contains a large number of massive stars, and it was postulated by Adriaan Blaauw in 1961, following a suggestion by Fritz Zwicky, that these three stars were in a quadruple star system, the fourth member of which exploded as a supernova flinging the three runaway stars far from their birthplace. The supernova remnant of the fourth star has long since dissipated.

Many astronomers believe this to be the general explanation for the massive runaway stars. It is also true that the runaway stars themselves will eventually undergo a supernova explosion, turning them into pulsars like their former companions, and they will then continue in their high-speed flight beyond the plane of the Galaxy. The Crab Nebula pulsar has a considerable proper motion away from the constellation Gemini and it may be that the supernova of 1054 occurred on a runaway star ejected about 4 million years ago from the I Geminorum association of massive stars.

It is not inconceivable that the momentum of the runaway stars and pulsars will in some cases be sufficient to carry them right out of the Galaxy. If the speed of the pulsars near the Sun exceeds some 290 km/s they have more than the velocity of escape from the Galaxy and will escape from it into intergalactic space. Pulsar number 0283+36 probably has such a speed and may be doomed to become an intergalactic tramp.

Presumably not every binary star system in which a supernova explosion takes place is disrupted. Sometimes the amount of material ejected will be less than half the total mass of the double star system so the laws of physics dictate that the pair will not break apart.

Perhaps the explosion itself will cause the pulsar formed to recoil in such a way that it remains in orbit around its companion. Why is it then that binary pulsars are so rare? The answer is presumably that the huge envelope of ionized gas from the ordinary star surrounds the pair and traps any detectable radio emission from the pulsar. Such a cloud of gas is blown off by our own Sun, and is known as the solar wind. However, the pulsar does ultimately become detectable – as an X-ray pulsar.

Why a pulsar shines X-rays

In the course of time, the hydrogen fuel in the centre of the companion of an unseen pulsar in a binary star system gives out and the star begins to expand into a red giant (as outlined in Chapter 9). It may be that the star and pulsar are sufficiently close that before the star's growth to a red giant is complete its atmosphere begins to leak onto the pulsar.

It is easy to see that at a certain point between two stars their gravitational forces exactly cancel each other, so that an atom placed at this point might find it difficult to decide to which star it should fall. Jules Verne wrote a story in which the crew of a giant shell fired into space suddenly fell from the floor to the ceiling of their capsule as they passed the equivalent point between Earth and Moon. (He did not understand the concept of free fall, and that astronauts are weightless except when their space capsule is being propelled). The real situation in a binary star is complicated by the presence of centrifugal forces caused by the revolution of the two stars in orbit around each other, but nonetheless the gravitational field of each star has its own zone of influence within which all material belongs to that star. These volumes are teardrop-shaped with the points touching, and are called Roche lobes. When one star fills its Roche lobe its atmosphere may protrude beyond its lobe into the adjacent one and fall towards the other star. So gas is transferred from the putative red giant to the pulsar.

Now, the pulsar is small: it is a neutron star, some 20 km in diameter. When the gas from its companion falls upon the neutron star, the gas is compressed by the intense gravitational force. Just as compressing the air in a bicycle pump heats it, so the in-falling gas is heated, to a temperature which may be tens of millions of degrees. The hotter a body the more energetic (shorter wavelength) the radiation it emits. This gas is so hot that it shines not by emitting infrared radiation or light but by emitting X-rays. The neutron star becomes detectable as an X-ray star.

The power available from the gravitational field of a neutron star is enormous. A marshmallow falling on to the surface of a neutron star would explode with the energy release of a World War II atomic bomb.

Many X-ray binary stars are now known from observations by X-ray survey telescopes on board artificial satellites, particularly Uhuru, launched in 1970, the Copernicus satellite launched in 1972, and the Einstein satellite launched in 1979.

How do astronomers know that a particular X-ray source is part of a binary star? Take as an example the strongest X-ray source in Hercules, called Her x-1. This source switches off for 6.5 hours every 1.7 days. Similar behaviour is found among ordinary stars, too, the explanation being that they are double stars of the type called *eclipsing binaries*. Quite simply, one star hides the other for a while during its orbit because the plane of their orbits is roughly in our line of sight.

In the case of Her x-1, the companion was known to be a variable, called HZ Herculis, before the X-rays were detected coming from the source. It is variable because the X-rays shining from the neutron star heat one side, this side turning towards and away from Earth with the orbital period of the neutron star. Her x-1 itself is a pulsar, pulsing X-rays with a period of nearly 1.25 seconds. As the pulsar orbits HZ Herculis, the Doppler shift of its pulsing frequency can clearly be detected. In fact, the Doppler shift of the companion star can also be measured, since some of the pulsed X-rays are intercepted by the companion star where they heat up its surface and are re-emitted as pulsing visible light.

A sustained and difficult monitoring of this weak pulsation has been made by Berkeley astronomers Jerry Nelson and John Middleditch so that they have been able to measure the Doppler shift of the light pulses emitted from the companion star and determine its orbit about the neutron star.

From these measurements, astronomers can tell at exactly how many kilometres a second each star is travelling. Combining this with the orbital period of 1.7 days and the fact that the orbits are in our line of sight, all details of the orbits are known. The importance of this is that the masses of the two objects can be worked out, using the same orbital laws that Kepler deduced from observations of the solar system.

At last, astronomers have been able to get to grips with observational facts about a neutron star. Theoreticians had calculated that no neutron star could have a mass greater than twice that of the Sun. In vindication of their work, the mass of Her x-1 turned out to be 1.3 solar masses.

Unlike radio pulsars which are all slowing down, the best-studied X-ray pulsars, like Her x-1 and Cen x-3, are speeding up. This is probably because X-ray pulsars are members of binary systems by their very nature. Impact of the infalling matter on to the surface of the neutron star gives an impulse to the star, speeding it up.

Binary pulsars

In contrast to the average population of stars, where the majority is in multiple star systems, and of galactic X-ray sources, which are virtually all binary star systems, only 4 of the 330 pulsars known are members of double stars. One is in a very eccentric and elongated orbit about its stellar companion, suggesting that, when it formed in a supernova explosion, it only just failed to be severed from the double star system. It was the first-known binary pulsar, discovered in 1974 by Taylor and Hulse with the Arecibo telescope. It is called PSR 1913+16 (its position in the sky). Its period of pulsation is 0.059 seconds and it orbits an unseen companion with a period of 27 908 seconds (almost 8 hours). The characteristics of its orbit can be measured with great accuracy, since the pulsar provides a near-perfect clock whose Doppler shift can be well measured. The orbit's eccentricity is 0.62; this means that the pulsar moves in an ellipse whose minor diameter is 0.38 times its major diameter – the axes are nearly in the ratio of 2.5:1. The separation of the two stars is about the same as the diameter of the Sun. Even though the neutron star has a negligible

diameter, this means that the unseen companion can hardly be a normal star, as it would not fit in the pulsar's orbit. It is probably a white dwarf.

The closeness of the two stars and the strength of their mutual interaction means that the gravitational field between them is strong. The normal theory of gravity, as formulated by Newton, is not good enough to describe the binary star's motion: Kepler's laws, for instance, do not apply except as an approximation. General Relativity has to be used, as formulated by Albert Einstein. One of the most famous historical predictions of General Relativity is of the advancement of the perihelion of the orbit of a close planet. In effect, the planet describes an orbit which has the shape of a rosette, rather than of an ellipse, and the major axis of the orbit moves slowly around the Sun in the direction of motion of the planet: the perihelion (closest point of approach to the Sun) advances. Mercury is the planet closest to the Sun, and General Relativistic effects show most clearly in its orbit. The perihelion of Mercury does indeed advance in its orbit, but for decades astronomers have argued over whether the observed rate of advancement is of the size predicted by Einstein's theory. General Relativity predicts an advance of 43 arc seconds per century. This small amount is confused by additional effects which are the result of the oblateness of the Sun – it is squashed slightly at its poles, like a tangerine or satsuma. The observed periastron advance of the binary pulsar's orbit is 4.2 degrees per year (35 000 times more than Mercury's!) and, since the companion star is a compact white dwarf, it is nearly spherical; in any case, oblateness has less of an effect since the star is small. The observed rate of advancement of the periastron of the binary pulsar's orbit is in spectacularly good agreement with Einstein's theory.

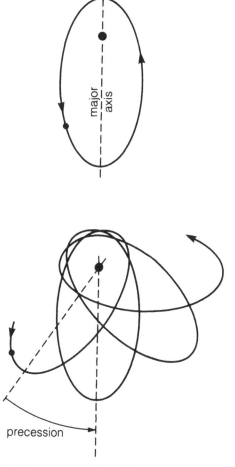

major axis

precession

FIG. 66. *Advancement of perihelion. 1. If Newton's theory of gravity is correct and two stars orbiting each other are small, of low mass and widely separated, then the relative orbit is an ellipse whose major axis of symmetry remains fixed in space. 2. If however the stars are massive or very close, and Einstein's theory is correct, then the elliptical orbit precesses, with the major axis rotating in space at a certain rate, and the orbit forming a rosette-like pattern.*

A radio binary pulsar was discovered in 1983 while Italian astronomers were investigating gamma ray sources, trying to identify the places from which the gamma rays came. They found a 6.1 millisecond pulsar (PSR 1953+29) orbiting an unseen companion with a circular orbit with period of about 120 days. Its companion is considerably less massive than the Sun and could be a white dwarf as in PSR 1913+16. It is not yet sure that the pulsar is the gamma ray source, and it certainly emits no X-ray emission. It is reminiscent of PSR 0820+02, which has a pulsation period of 0.86 seconds and an orbital period of 1223 days. It, too, has no measurable eccentricity.

To explain why these binary pulsars have circular orbits, astronomers conjecture that they are not recently formed pulsars. Maybe they are the relics of previous X-ray binary systems. After an initial supernova explosion which formed an eccentric neutron star binary system, the primary star became a giant. It transferred mass onto the neutron star, causing it to become an X-ray star. The accreted mass pushed the neutron star into a circular orbit. Eventually the primary star became a white dwarf, and the circular orbit remained.

The Millisecond Pulsar

Until 1982, the Crab pulsar was the fastest pulsar known, but in that year a radio source known as 4C21.53 was shown to be an even faster one. The radio source had been an enigma for many years and the tracking down of its true nature was a classic piece of astronomical detective work by D. C. Backer, Shrivinas Kulkarni, Carl Heiles, M. Davis and Miller Goss.

4C21.53 showed interplanetary radio scintillations like those in quasars, for the study of which Hewish had built the 4½ Acre Telescope in Cambridge, with which he and Bell discovered the

Crab pulsar. The scintillations indicated that the radio source was of small angular size, but its position near the galactic plane contradicted the idea that it was a distant quasar. As radio waves from quasars travel the length of the galactic plane, they are refracted by the electrons in interstellar clouds and the quasars' images are broadened, like street lights seen through a misty windscreen. Quasars in the galactic plane thus appear to be of larger angular extent than they really are, and scintillations produced by interplanetary clouds are consequently of less effect. Thus if 4C21.53 showed strong interplanetary scintillations, its image could not have been broadened by interstellar clouds and it must be much nearer than a quasar, a galactic rather than an extragalactic object.

Secondly, the point-like object at first seemed superimposed upon an extended non-thermal radio source. The object had the radio appearance of the Crab Nebula, with the implication that the small object was a pulsar like the Crab pulsar. This caused astronomers at the Arecibo and Owens Valley Radio observatories to search, in 1979, for a pulsar in 4C21.53. They had no success. Their search technique would have found sufficiently bright pulsars with periods longer than 10 milliseconds. Faster than this, the pulses would have smeared into one another.

It was then discovered that, as occasionally happens with an interferometer, the extended 4C radio source had been mislocated to the west of its true position. The view by a radio interferometer of a radio source shows multiple peaks, with the strongest peak representing the position of the object. If, however, the smaller so-called side-lobes are confused with other sources (as they were in this crowded field), then radio astronomers may misidentify the strongest peak and place the source one or two lobes east or

west of its true position. This had happened with 4C21.53 and an eastern source (now known as 4C21.53E) had been superimposed on the scintillating source in error. 4C21.53E is indeed an extended non-thermal radio source – but it has since been shown to be a conventional double radio source and is a radio galaxy. While the extended non-thermal radio source was thus taken out of the picture it still left the problem of the nature of 4C21.53W.

A further confusion then had to be sorted out. Two measurements of 4C21.53W gave different answers for its position. One was made at the Culgoora interferometer in Australia, which senses radio waves of 80 MHz frequency. The other was made at Bonn with radio waves of 5000 MHz frequency. The position found at Culgoora was south of that found at Bonn. This suggested that 4C21.53W was itself double, with each source in turn dominating at the different frequencies. A radio map made at an intermediate frequency (609 MHz) by the Westerbork Synthesis Telescope in the Netherlands confirmed this: the point-like object lay south of a small extended thermal radio source to the north. This nebula was, however, not a radio supernova remnant, but an ordinary nebula probably in the line of sight by chance.

Armed with this information, astronomers carried out another search for a pulsar at the position of 4C21.53W at Arecibo in March 1982. The equipment had been improved so that it would have been sensitive to pulsars with periods as short as 4 milliseconds. Any reasonable person would have expected that a search sensitive to pulsars with periods almost 10 times shorter than the fastest pulsar then known would have had a good chance of success, but no pulsar was found. In September 1982, the Westerbork Synthesis Telescope measured strong polarization in

4C21.53W; this made the existence of a pulsar there all the more likely, since strong radio polarization is associated with strong magnetic fields, of the kind that are found in rotating neutron stars. The only possibility was that the pulsar was of a period even shorter than 4 milliseconds. The equipment was further improved, and within 1 month the team had discovered a pulsar in 4C21.53W with a period of 1.5 milliseconds.

The pulsar, whose name is PSR 1937+214, is also called the Millisecond Pulsar. Its period has been accurately measured as 0.00155780644885 seconds. No change has been detected in its period; it is almost as accurate as the most accurate terrestrial clocks known. Indeed, if irregularities are found it will not be obvious whether they are in the pulsar or in the clocks used to time it. It is astonishing that the fastest pulsar known shows no changes in its period, since astronomers would have expected that the fastest pulsar would be the youngest, radiate most energy and spin down the quickest. The spin-down rate is in fact less than 1 million millionth of a second per year.

Explaining how this pulsar formed is a problem. There is no supernova remnant nearby. This means that the pulsar is more than 10 000 years old; the slow spin-down rate implies that it is more than 1 billion years old. Why has it not slowed down? Alternatively, if it is old and has at one time slowed down, what has speeded it up again? Some theories look to the formation of a pulsar with a weak magnetic field, and therefore poor brakes. Such a pulsar would spin for a very long time indeed. Other theories suggest that the pulsar was a neutron star formed in what became an X-ray binary system. It accreted mass and angular momentum and became much faster. Then it lost its companion in a supernova explosion.

Pulsars as musical notes

The fastest pulsars are listed in Table 5. Their periods range from 1.6 to 89 milliseconds. The frequency of vibration of a pulsar is the number of pulses which it emits per second. If the pulses were converted to sound waves of the same frequency then they would be heard as notes. The notes of the equal tempered chromatic scale which correspond to each pulsar frequency are given in the table. On the notation used, middle C on a piano is C4. E5 is the E above high C, the ninth white key to the right of middle C (not counting middle C itself). E3 is the E below middle C, the fifth white key left of middle C, and B0 is 22 to the left, among the bass notes. C#1 is the left of the pair of black keys to the low end of a piano keyboard (between white keys 20 and 21 to the left of middle C); C#0 is one octave below that, off the scale of pianos. C#0 is, in fact, the sharp of the lowest C on an organ, played by an organ pipe nearly 5 metres long, and about the lowest note which the human ear can hear. The Vela Pulsar is at too low a frequency to be represented in musical notation.

An X-ray binary in a supernova remnant

Among the known X-ray binary stars, one, with catalogue number 1E2259+586, is remarkable because it was discovered in a supernova remnant CTB109. The supernova remnant had been known since 1960: it lies close to Cas A, which is so intense that the study of CTB109 is difficult, just as the faint star Sirius B is hard to see in the glare of Sirius itself. New radio techniques were brought to bear on CTB109 in the 1980s and it shows a semicircular arc of radio emission. There are some faint optical filaments nearby. The distance of CTB109 is 10 000 light years and the supernova explosion which formed it occurred

Table 5.   *The fastest pulsars*

	Period (s)	Frequency (Hz)	Note
The Millisecond Pulsar	0.0016	642	E5
Binary Pulsar PSR1953+29	0.0061	164	E3
Crab Pulsar	0.033	30	B0
LMC Pulsar	0.050	20	E0
Binary Pulsar PSR1913+16	0.059	17	C#0
Vela Pulsar	0.089	11	—

about 10 000 years ago. When Canadian astronomers Phil Gregory and G. C. Fahlman used the Einstein Satellite to look at CTB109 they were surprised to discover in the centre of the supernova remnant an X-ray pulsar, 1E2259+586. Its pulse period is 7 seconds and there are indications that it orbits its companion with a period of about 1 hour; this would imply that the two stars in the double system are rather small – a neutron star and white dwarf, for instance. The neutron star would then have to be accreting mass from the common envelope which surrounds the two stars, rather than by mass transfer from one star to the other. The orbit of the neutron star is eccentric.

In the centre of the arc of the supernova remnant, surrounding the position of the X-ray pulsar, there is radio and X-ray emission. This seems to be from material ejected by the binary system, perhaps by the pulsar itself.

The story behind the X-ray binary and CTB109 is puzzling, and astronomers are not sure of the evolutionary history of the system. Did they start as a pair of white dwarfs? In this case, a rather small mass was ejected from one by an unknown mechanism. Was it once a white dwarf and an ordinary star which went supernova? In this case, the star must have been very small to fit

into the available space in the neutron star orbit. Why did such a small star turn into a supernova? Possibly this whole object is a chance coincidence between an X-ray star and a supernova remnant, but the central emission argues against this idea.

In spite of the questions it is gratifying to see the correlation of a supernova remnant with an X-ray binary star. There are two possible further examples, although it is not actually known whether the X-ray stars which lie at the centres of the supernova remnants concerned are indeed binaries. The supernova remnants are called Kes 73 and PKS1209-52. With CTB109 they add three probable examples to the list of connections between neutron stars and supernovae.

Type I supernovae

Type II supernovae may occur, almost by accident, in binary star systems, and cause the stars to recoil, disrupt or become eccentric. Type I supernovae, on the other hand, appear actually to be caused by being part of a binary star system. The crucial piece of evidence is the absence of hydrogen from the spectra of Type I supernovae. This suggests that a very evolved star is the progenitor of Type Is – a star which has already lost its outer layers of hydrogen (by a stellar wind, for instance). Astronomers suspected that a white dwarf or similar star is involved in the production of a Type I supernova. White dwarfs can have only up to about 1.5 solar masses, and the fact that astronomers at the Einstein Observatory calculated that the mass of ejecta in Tycho's supernova remnant (a definite Type I, as we saw in Chapter 8) was about 1.9 solar masses encouraged this speculation (as already mentioned, the calculation of the mass depends critically on the distance assumed for Tycho's supernova, and 1.9 is not different from 1.5 in this context!).

Another clue about the mass of Type I supernovae came from the light curves of Type Is. They were so uniform in their decay that they were reminiscent of the decay of a radioactive element. In 1962 T. Pankey of Harvard University proposed nickel-56 as the element responsible for this decay. This was attractive, since the spectra of Type I supernovae showed bumps which could be identified as combinations of iron line emission – and nickel-56 eventually decays to iron. The mass of nickel-56 which is required to generate the energy output of a Type I supernova is about 1 solar mass. This was another hint that white dwarfs were involved.

The fact that white dwarfs are unstable if their mass is larger than the Chandrasekhar Limit of about 1.5 solar masses suggests that Type Is can be made if a white dwarf is pushed over the limit. How could this happen? Craig Wheeler of the University of Texas says:

You place a white dwarf in a close binary and you dump mass on to it.

If that is the basic scenario, there remains a large number of variables in the theory for astrophysicists to play with; this is what they have been doing in the last decade.

As each theoretician calculates the result of his own choice of variables, gambling on his experience and his nose for what is likely to be correct, a large number of models of Type I supernovae accumulate. Most fail and are discarded when confronted by some unforeseen problem; the ones left are those which are nearest the truth and which become accepted as The Theory of Type I supernovae.

The basic idea is that the white dwarf is orbiting an ordinary star which evolves towards the red giant stage and leaks hydrogen on to the white dwarf. A variant of this is that the white dwarf has a companion which has itself become a white dwarf in the past. If the two are so close that they have a common atmosphere, then helium from one may leak on to the other and bring it over the limit.

The second variable is the composition of the exploding white dwarf. It is the core of a heavy star whose outer parts have been stripped away by a stellar wind. Depending on the size of the star from which the white dwarf evolved, the core could be helium, or a blend of carbon and oxygen, or a mixture of heavier elements. The most popular models depend on properties of carbon burning, so most theoreticians choose the carbon/oxygen white dwarf as the kind most likely to explode.

As the white dwarf accretes mass from its companion, the extra weight increases the inward pressure at its surface; this is compensated by a rise of density and temperature in the central part of the white dwarf. Eventually the internal temperature reaches 4000 million degrees, at which the carbon ignites and burns in a nuclear explosion. The release of energy increases the central temperature even more, but the white dwarf cannot adjust to this by expanding like a normal gaseous star, because it is degenerate. So the carbon burning spreads rapidly through the

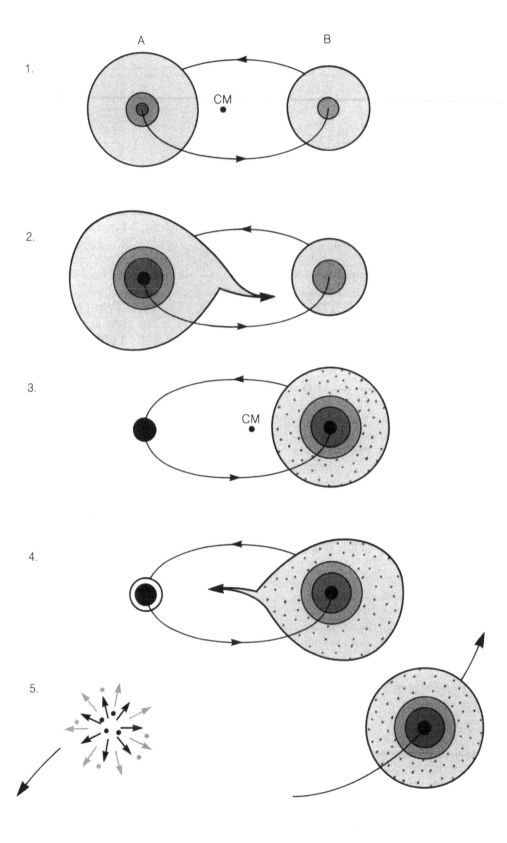

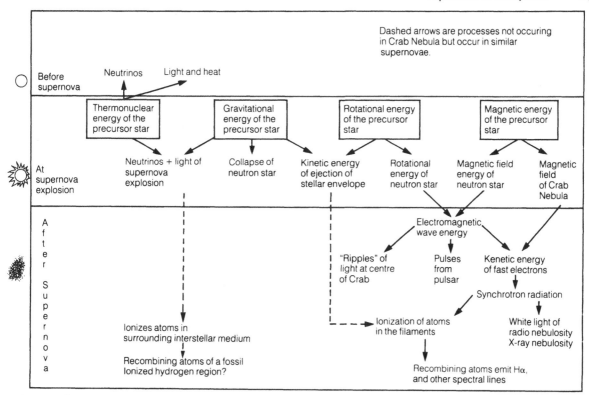

FIG. 68. *(Above) Energy flow in a supernova.*

FIG. 67. *(Left) Supernova of Type I. Five successive stages lead to the detonation of a carbon-oxygen white dwarf. 1. The sequence starts with a double star in which the more massive, A, is burning hydrogen in a shell around a helium core. 2. A becomes a red giant with a carbon-oxygen core produced by helium burning. Its hydrogen and helium envelopes expand and overflow onto the less massive star B. 3. Eventually, A is reduced to its carbon-oxygen core and it circles in orbit around B, now the more massive. B has a dirty envelope enriched by A's nuclear processed material. B follows much the same nuclear history as A. 4. As it, in its turn, becomes a red giant, B overflows back to A. 5. A detonates and mostly disintegrates, releasing B like a sling shot to become a runaway star and, eventually a pulsar.*

star. It first produces oxygen, neon and magnesium and then progresses to nickel-56.

This is the point where astrophysicists choose a further variable in the white dwarf model of the Type I supernova. They might allow the carbon burning to spread quickly through the white dwarf – in less than 1 second! – and this explosion is called *carbon detonation*. The whole of the white dwarf is suddenly converted to nickel-56 with the release of huge amounts of energy. This causes the outer layers of the white dwarf to be ejected at 30 000 km/s, leaving behind a small white dwarf which is made of nickel-56 which eventually decays to iron.

Alternatively, in the *carbon deflagration* model, due principally to Tokyo astrophysicist Ken'ichi Nomoto, the carbon burning spreads more slowly through the white dwarf – in 3 seconds! This produces about half the quantity of nickel-56 and a range of other products like calcium, sulphur and magnesium; it completely disintegrates the white dwarf. This model is attractive since the spectra of Type I supernovae can be interpreted as containing spectral-lines due to elements like these and (as seems to be the case for Type Is) no stellar remnant is left behind.

In either case, however, the energy release (which is initially the energy of the thermal runaway of carbon burning) is continued by radioactive decay of the nickel to cobalt and then to iron. The slower of these decays, which sets the pace of the whole process – just as the pace of a convoy is set by the slowest ship – is the 77 day half-life of cobalt-56. This could match the 60 day half-life of the Type I supernova light curve; one has to take account of the effect that radiation inside the supernova finds it progressively easier to leak out of the expanding debris, as it thins, speeding the light curve's decay. The details of this have, however, not been made to fit the facts. Some astrophysicists warn against jumping on the bandwaggon of the white dwarf binary theory. Even if the theory is correct in broad outline, it is too early to pick the best set of variables described above and, indeed, Icko Iben of the University of Illinois leaves open the possibility that there is more than one evolutionary scenario that culminates in the production of a Type I supernova.

The energy flow in a supernova

The energy budget of a supernova starts with four kinds of energy which it has 'in the bank'. The precursor star has thermonuclear energy by which it shines; it also has gravitational, rotational and magnetic energy. At the supernova explosion, the thermonuclear and gravitational energy are let loose to produce the flash of light and other forms of radiation by which we sense the supernova and which ionizes the surrounding interstellar medium to produce a fossil nebula. The kinetic energy of the ejection of the stellar envelope causes thermal X-rays and ionization of the shock material in the surrounding interstellar medium. The rotational energy of the precursor and its magnetic energy combine to become the electromagnetic wave generator which produces all the phenomena associated with a pulsar. The flow of energy forms a network which is illustrated in Fig. 68.

11

Creation of the elements

Act first, this Earth, a stage so gloomed with woe
You all but sicken at the shifting scenes.
And yet be patient. Our Playwright may show
In some fifth Act what this wild drama means.

TENNYSON

In some of the most rugged mountainous country of New South Wales, along the Turon River and in its hinterland, are the ghost towns of the Australian gold rush of the 1870s. A few people are left in Hill End, but of the town of Tambaroora nothing remains except Golden Gully, a canyon dug by miners in their thousands. Tens of millions of tons of dirt were dug from here, washed and scrutinized for the glint of a metal: precious because both beautiful and rare, and it is rare because of the rarity of supernovae.

Abundance of the elements

Lured to inhospitable terrain because of the incredible richness of the strike, those miners who were luckier than most took from the goldfield a total of just 20 tons of gold, on average 1 gram for every ton of dirt dug over. Modern gold mines operate profitably when there is a worthwhile concentration of gold of about 20 grams (about an ounce), per tonne in contrast with the average concentration of gold in the surface of the Earth of approximately one thousandth of a gram per ton.

What of the concentration of gold averaged over the entire mass of the Earth? There is, of course, no direct evidence since the central regions of the Earth are inaccessible. However, it is a speculation commonly held by astronomers that meteorites – stones and rocks which have fallen from space to the Earth – represent the remains of a defunct planet as solid as the Earth or Moon, so that the abundance of the elements in meteorites

may be like the abundances in the interior of the Earth. Gold is 100–200 times more abundant in meteorites than in the surface of the Earth, so is presumably similarly more abundant in the Earth's centre.

It is not difficult to guess why this should be. The Earth's interior is hot, because of heating by radioactive materials, and partly perhaps because of its contraction under the force of gravity. In some ways it resembles an ore-smelting furnace, melting rocks so that the lighter slag rises to the surface to make the Earth's crust, while the metals such as iron, nickel and gold fall towards the Earth's centre.

Direct determining of the abundance of gold in stars is not possible. Only one clear spectral line in the Sun's spectrum caused by atoms of gold has been found, proving gold's existence there, but since the mechanism in the gold atom that causes the spectral line has not been studied well enough, no accurate estimate of the amount of gold required to form the line can be inferred. Its concentration in the Sun has been estimated by assuming that, since nickel and gold have similar properties, the ratio of their concentration in meteorites is the same as in the Sun. Because nickel gives rise to many well-studied spectral lines in the solar spectrum, its abundance can be measured easily.

Approximately 1 milligram in every ton of the Sun is gold. Although it is not possible to identify gold as such in the spectra of stars, iron lines are relatively common; broadly speaking, we can prorate the gold concentration in stars with the iron concentration. On this basis, the Sun would have a 30 000 times greater concentration of gold than would be found in these stars with the lowest gold (i.e. iron) concentration.

Why is gold rare? Why is iron, on the other

FIG. 69. *Golden Gully, Tambaroora, New South Wales. Millions of tons of dirt were dug by pick and shovel from this gully to collect gold, grain by grain. The rarity of gold is linked to its formation deep inside massive stars. Because supernovae, scattering their heavy elements through space, are rare, gold is rare too.*

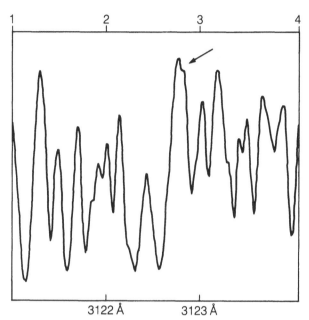

FIG. 70. *Gold in the Sun. The valleys in this graph of 0.03% of visible spectrum of the Sun represent chunks of colour 'bitten' from sunlight by atoms, mostly of iron, in the Sun's atmosphere. Peaks represent colours which have passed relatively freely out of the Sun. The nibble arrowed is the only known evidence for gold in the Sun. It hints at the wave of supernovae which manufactured the gold, and enriched the pre-solar nebula from which the Sun formed.*

hand, relatively common? Why is it that in spite of the great diversity of astronomical objects whose composition has been studied – the Earth, meteorites, the Sun, most stars – the relative abundances of the elements in all these bodies are surprisingly similar, and the differences are readily explainable by some plausible guesses and the histories of the bodies?

Clearly there is some common astronomical explanation for the origin of the elements in all these celestial objects, and supernovae play a crucial role.

The alpha-beta-gamma theory

Modern discussions have a two-pronged attack on the creation of the elements. The first starts, as the Universe started, with the Big Bang. This theory of the origin of the elements was originally called the Alpha-Beta-Gamma theory, in part because the elements are supposed to be formed in sequence like the start of the Greek alphabet, and in part because the theory was proposed in detailed form by Ralph A. Alpher, Hans Bethe and George Gamow in 1948. (Bethe's part in creating the theory was small – Gamow said that he invited Bethe to be a co-author because the pun appealed to Gamow's sense of humour.) Gamow called the material of the Big Bang *ylem*: he envisaged it as a gas made of neutrons, although modern authors see it as a more complicated mixture. When the theory was first put forward it was to account for the formation of all the elements from this basic ylem. As we shall see, however, it would complete only the first stage of the process.

The neutrons in ylem changed relatively slowly into protons and electrons as the Universe got under way. Some of the protons thus formed captured neutrons to make more complicated nuclei. Some of these nuclei would change by beta-decay (emission of an electron) and some would gather further neutrons to become more complex nuclei. All the element building in the Universe, according to the alpha-beta-gamma theory, occurred in the first 2 hours that the Universe existed. As the Universe cooled, the nuclei would capture the free electrons to become atoms.

One feature of the abundance of the elements which this theory explains well is the fact that nuclei whose ability to capture neutrons is low are more common than those whose ability is high.

Table 6. *Abundance of gold*

Location	Concentration (gram/ton)
Profitable gold mine	20
Gold strike, with many unsuccessful mines	1
Interior of Earth	0.06
Surface crust of the Earth	0.001
Sun	0.001
Old stars	0.00000003

Think of nuclei as a large intake of graduates into a big organization in which promotion is by merit. Bright graduates (nuclei with high neutron-capture ability) are susceptible to promotions; they take advantage of random opportunities (neutrons) as they occur and are promoted faster than their fellows (they form more complex nuclei). The duller graduates' ability is lower, they stay in their career grades longer and consequently there are more of them than of the bright graduates.

Among the elements certain nuclei have exceptionally low ability to capture neutrons – these occur at the so-called magic neutron numbers 50, 82 and 126. Elements occurring at these numbers, like lead, are exceptionally common.

The alpha-beta-gamma theory suffers, however, from a fatal flaw. It requires that the nuclei build up from hydrogen by adding one neutron at a time, with the neutron changing to a proton at appropriate stages. The flaw is that at two vital stages, the nuclei created cannot exist for more than a minuscule fraction of a second (a thousand, million, million, millionth of a second!) after which time they release the neutron that they have just captured and return to a helium nucleus again. They do not exist for long enough for the

next step to occur. The break in the chain occurs just after the formation of helium, about 2 minutes after the start of the Big Bang. Thus the alpha-beta-gamma process cannot proceed past helium in making the elements. No further, heavier elements can be made. If the alpha-beta-gamma process of element building were all that could occur, the Universe would consist solely of hydrogen and helium and there would be no carbon, no nitrogen or oxygen, no paper to make this book, no writer to write it, no reader to read it.

Creating elements inside stars

There must have been other sites in the Universe, apart from at its beginning, where the heavier elements were created. When the process by which nuclear energy became starlight was discovered as a result of the work by Bethe and von Weizsäcker in 1939, it was realized that the same processes would change the composition of the stars and create new elements.

The first observational evidence that the stars create elements was Paul Merrill's discovery in 1952 of spectral lines of the element technetium in red giant stars. Technetium is unstable on an astronomical time scale and lasts at most for a few million years. As the red giant stars were

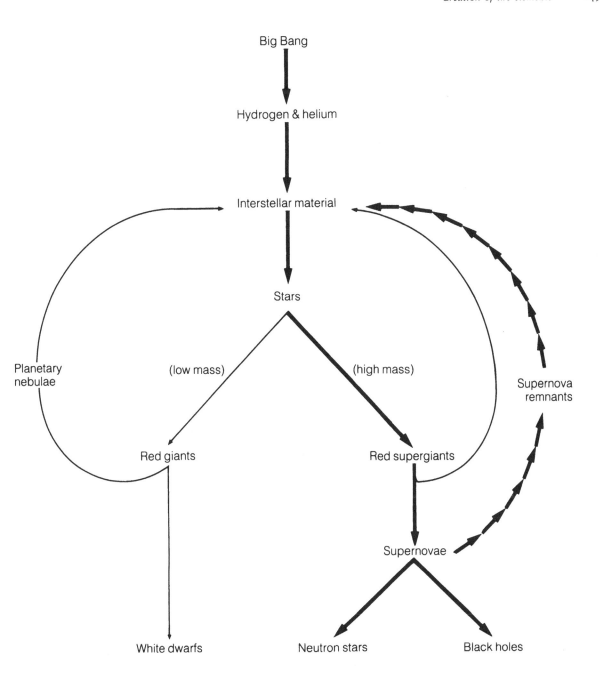

FIG. 71. *Chemical evolution of the Galaxy. The flow is from the Big Bang which created hydrogen and helium from which stars form. Gas is recycled back into the interstellar medium via planetary nebulae or supernova remnants, or ejected by low surface gravity supergiants, but everything ultimately ends as white dwarf, neutron star or black hole.*

known to be older than this, clearly technetium could not have been in these stars since they were formed, it must have been made there. Other stars were discovered which, like red giants, had developed sufficiently in their evolutionary life to show an excess of carbon or oxygen caused by helium burning in the so-called triple-alpha process.

The time was ripe for a detailed examination of the formation of the elements in stars. The problem engaged the attention of Fred Hoyle in 1946 because of the advocacy of the Steady State Theory of the Universe in which there was no Big Bang and therefore no cosmological element creation. Hoyle was faced with the existence of the wide variety of elements which he had to explain in other ways, and one of the main successes of the Steady State Theory was to stimulate this research, although the theory has since lost credibility as a cosmology. The specific processes which form the elements in stars were detailed in a foundation-laying paper in 1957 by Geoffrey Burbidge, Margaret Burbidge, Willy Fowler and Fred Hoyle, known as the B^2FH (B-squared, F, H) paper. Fowler received in 1984 the Nobel Prize for his work on nucleosynthesis, and thereby the Nobel Prize committee created another scandal. Why did the other authors, particularly Fred Hoyle, not receive a nomination? Whatever the Nobel Prize committee thought, the 1957 paper was a joint effort, in which B^2FH supposed that the first stars consisted principally of hydrogen. Most stars visible now are in the process of converting that hydrogen to helium, releasing energy which can be seen as starlight.

This process creates helium from hydrogen. As stars age, they 'burn' some of the helium which they have created to produce carbon, oxygen, neon and magnesium. The interior of the star may be mixed, stirred up by clouds of hot material billowing from the star's centre towards its surface by the force of convection. Depending on whether mixing occurs or not, the carbon and oxygen can capture protons (hydrogen nuclei) or alpha particles (helium nuclei) to make either proton-rich nuclei or the elements magnesium, silicon, sulphur, argon and calcium.

B^2FH identified five further processes occurring in or on the stars. The first, which they named the *e*-process, occurs somewhere in some kind of star yet to be satisfactorily identified. When the mixture of elements formed in the previous processes cooks at a high temperature, the protons in the mixture absorb electrons and release them at equal rates in a situation of *equilibrium* (hence the name *e*-process). This creates those abundant elements such as iron, nickel, chromium and cobalt which are known collectively as the iron peak.

Up to this point the element-building process has been relatively easy: the creation of heavier elements has released energy. In a sense, the nuclei of these elements *want* to be formed (in the same sense that a heated object *wants* to cool by radiating energy). Beyond the iron peak however, the processes creating heavier elements have to have a supply of energy available to do so. Such a supply is available in a supernova explosion. Just before the supernova, the precursor star has built up a supply of middleweight elements by burning helium and carbon. Something happens and the precursor becomes unstable: the precise way in which the supernova occurs may not be important from the point of view of the creation of heavy elements. What matters is the speed at which the process occurs, so that neutrons are added to the middleweight nuclei sufficiently quickly that the successive nuclei, though unstable, do not have time to eject electrons and turn into something else. Because it is *rapid*, this process is called the *r*-process. (The *s*-process is one in which successive neutrons are *s*lowly added to middleweight nuclei which do have time to decay before the next neutron comes along.)

The *r*-process forms most of each of the following elements (in total about 1 Earth mass per supernova): selenium, bromine, krypton, rubidium, tellurium, iodine, xenon, europium, gadolinium, terbium, dysprosium, holmium, erbium, thulium, ytterbium, lutecium, rhenium, osmium, iridium, platinum, gold and uranium. The reader may find the names of some of these elements unfamiliar. Others he will recognize as precious because they are uncommon. The reason for their scarcity is the infrequency of supernovae.

A process called the *p*-process takes place when protons are added to nuclei formed by the *r* and *s* processes, but the *p*-process creates only a minority of the atoms of any one element. It may occur as the supernova ejects its outer layers into space, these layers containing unburned hydrogen and hence abundant protons, as well as *r*- and *s*-process nuclei.

Scattering the elements

The elements were made in stars. How do they come to be found in the Earth?

The elements created in stars, including those created in supernovae, are thrown into space by the supernovae themselves. Most of the progenitor star is ejected into space by the force of the supernova explosion. At first the supernova remnant is made of just the ejecta and we could expect to see peculiarities in the supernova remnant's composition – it consists of the insides of a star. Later the supernova remnant sweeps up ordinary material from interstellar space, a process which dilutes any peculiarities, as well as slowing the ejecta. Thus, to be able to see the dissected remains of stars, astronomers should

look at the young, fast-moving supernova remnants.

Several young supernova remnants show spectral peculiarities. The general interstellar medium, as represented by nebulae which have been heated by the nearby presence of bright hot stars, has a spectrum which has strong hydrogen lines – not surprising since 90% of the interstellar medium is hydrogen – and lines of helium, nitrogen, oxygen; lines of other elements are weakly present. By contrast, the fast-moving knots of Cas A have spectra which show very strong oxygen lines, and moderately strong sulphur, calcium and argon lines, but not a trace of hydrogen, helium or nitrogen. If Cas A was a Type II supernova in which a massive star of 20–60 solar masses had exploded, then this is understandable; the lighter elements in the star's outer envelope have been taken away in a stellar wind, and what we now see represents the exploded middle of the star, the shattered series of 'inner onion skins' of heavy elements.

A second supernova remnant like Cas A is known as G292+1.8; it has a similar oxygen spectrum and resulted from a similar star's explosion. This supernova is in the southern sky, and its remnant's speed and size suggests that it occurred about 1000 years ago. It undoubtedly would have been a historical supernova if there had been literate civilizations in Australia, South Africa or South America to record it. It (or the Vela supernova of 10000 years ago) may be the bright event depicted in Australian aboriginal pictographs called 'sunbursts'.

The Crab shows more subtle evidence for spectral peculiarities. It has a spectrum similar in appearance to spectra of the general interstellar medium, but the ratios of the spectral lines are not quite the same. The helium lines are stronger relative to the hydrogen lines, and, indeed, vary

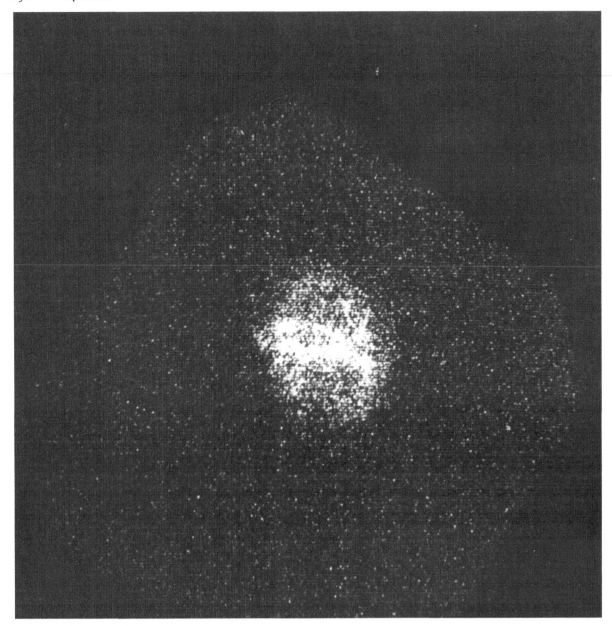

FIG. 72. *(Above) G292.0+1.8. A curious and unexplained pattern of X-ray emission lies inside an overall circular patch representing the supernova remnant G292.0+1.8. Some astronomers think that the pattern is a wavy ring connected with rotation of the star. This Einstein Satellite image was obtained by David H. Clark and Ian Tuohy.*

FIG. 73. *(Right) Sunburst. Unexplained rock carvings in the Australian desert near Broken Hill, New South Wales, are called 'sunbursts'. Perhaps they express a reaction by Australian aborigines to the bright and unexpected appearance of a southern hemisphere supernova, like the Vela supernova of some 10 000 years ago. Photo by Mike Daley, Australian Broadcasting Commission.*

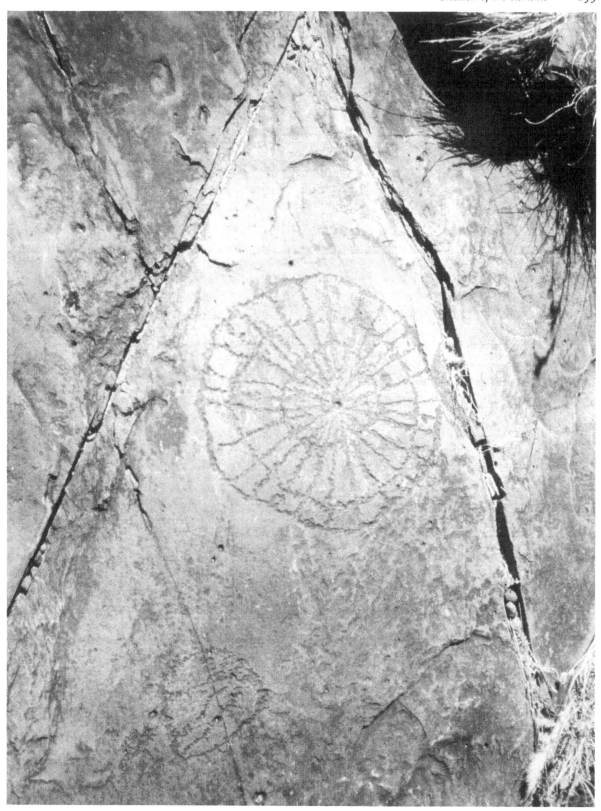

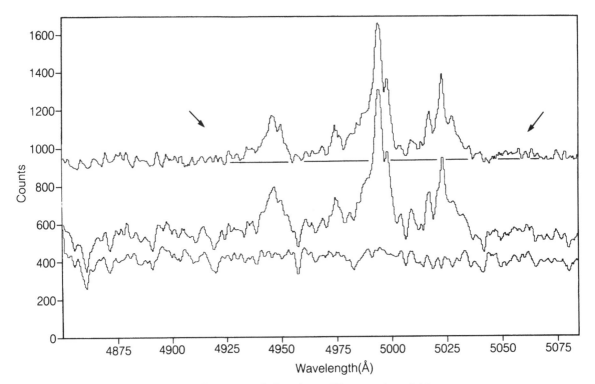

FIG. 74. *High velocity material in the Crab Nebula. The middle part plot of this spectrum of the Crab Nebula filaments shows a complex structure caused by the Doppler Effect acting on just two spectral lines. The spectral lines stand up as bright hills on a background of light from the synchrotron emission of the nebula and from faint Milky Way stars (bottom plot). The difference (upper plot) represents the spectrum of the filaments. Emission extends to the two arrows representing the fastest approaching and receding material respectively. It lies well in front of and behind the main body of the nebula, which is responsible for the main emission. Data by David H. Clark and Paul Murdin with the Anglo-Australian Telescope.*

Table 7. *Creation of the elements*

Process	Where?	Some of the elements formed
Alpha-beta-gamma process	Origin of Universe (Big Bang)	Hydrogen, helium
Hydrogen burning	Most stars	Helium
Helium burning (triple-alpha process)	Red giant stars	Carbon, oxygen
Carbon burning	Massive red supergiants	Neon, sodium, magnesium, silicon
e-process	Hot centres of stars? Supernovae?	Iron, nickel, chromium, cobalt
s-process	Evolved stars	Copper, zinc, lead, technetium
p-process	Surfaces or shells of supernovae	Small quantities of molybdenum, samarium
r-process	Supernovae	Gold, platinum, 'rare earths'
x-process	Surfaces of stars? Cosmic rays?	Lithium, beryllium, boron

from place to place in the nebula. The Crab seems to have been the result of an explosion of a star of about 9 solar masses. It was therefore a Type II, in accordance with the discrepant point on its light curve (Chapter 8). One difficulty with this notion is that astrophysicists calculate from the observed light from the Crab that it has a mass of about 2 solar masses. Add one or two solar masses for the mass of the neutron star pulsar, and the total sums to at most 4 solar masses. 5 solar masses are missing. There is some photographic evidence that the Crab is surrounded by a faint halo and there is some spectroscopic evidence that the Crab contains high-speed material which must, if it originated in the 1054 explosion, lie outside the main body of the nebula. This faint extension of the Crab could increase the mass of the Crab to about the 9 solar masses required. So far, however, the evidence is not completely convincing and the question as to whether the supernova of 1054 was really Type II is still unresolved.

Unfortunately, it is not possible to pick out in the spectra of supernova remnants the *r*-process elements. We have to wait about half a billion years after the supernova explosion to be able to obtain evidence for *r*-process material. The supernova ejecta take about 1 million years to merge with the general interstellar medium and get mixed up with all the rest of the material there. New stars condense from the interstellar material, shining first as large cool stars, not visible to optical astronomers but emitting enough infrared radiation to be detected by specialized infrared-detecting telescopes. Such stars spin fast. Having contracted from a much larger slowly rotating interstellar cloud, new stars spin up as they contract, just as a neutron star spins up as it is exploded from the centre of a supernova.

Nonetheless, because it is an observed fact that most stars spin slowly, newly formed stars must shed their angular speed by creating planetary systems such as ours. Virtually the whole of the rotational energy of the solar system lies in the massive planet Jupiter, which orbits the Sun once every 12 years. If the Sun had not ejected Jupiter its rotational period would have been less than about 3 hours instead of its present period of just over 25 days. The newly formed Sun therefore shed some of its material in what is known as the solar nebula to make the solar system; and that is how the *r*-process elements, like gold, travelled from a supernova in the distant past and came to be found in the Earth, to be sought and dug up by the miners of the gold rushes.

Cosmological clocks

How distant a past was it? The study of the age of the elements is somewhat grandly called nucleocosmochronology, and it exploits the fact that some of the atomic nuclei created in the *r*-process are not stable – they spontaneously decay into something else, perhaps on a very long time scale indeed. For example, the nucleus of the iodine atom exists in two long-lived forms, one of which is very stable and lasts indefinitely, while the other decays into a form of xenon gas on a time scale of 17 million years. Both kinds of iodine are produced in approximately equal amounts by the *r*-process in supernovae but, because one (so-called iodine-129), changes to xenon-129, the other (iodine-127) eventually predominates.

While all the *r*-process elements are floating about in some interstellar cloud; the xenon-129 produced by the iodine-129 dissipates into space. But after the *r*-process elements have condensed into planets and meteorites, the xenon-129 produced by the decay of iodine-129 is trapped in the rock containing the iodine. The amount of xenon-129 in the rock thus tells how much iodine-129 has decayed since the rock solidified.

We know that there were equal amounts of iodine-127 and iodine-129 when the *r*-process which formed them took place; we know how

much iodine-129 has converted into xenon-129 while the elements were in interstellar space. Because we know how long this takes, we now know the time that the *r*-process elements were in space before the rock solidified: some 200–600 million years (say half a billion years, in round numbers).

We have measured the time between the supernovae which formed the *r*-process elements which condensed into the solar system, and the formation of the solar system itself.

Is this a reasonable time? Stars form in the spiral arms of the Galaxy. There the dust and gas between existing stars is compressed, enabling new stars to condense from it. Spiral arms swirl around the Galaxy, causing gas at a given place in the Galaxy to be compressed repeatedly, the time interval between compressions being measured in hundreds of millions of years, the rotation period of the Galaxy. If the elements dispersed into the interstellar medium by supernovae occurring between successive compressions are all condensed out to stars at each compression, the average time these elements are free in space would be roughly half the interval between compressions. If, as is more likely, just a fraction is condensed at each compression, the average time the elements are free in space would be longer, but would presumably still be measured in hundreds of millions of years – just about the length of time actually determined by nucleocosmochronology.

Nucleocosmochronologists (if, indeed, there are any astronomers who put up with the inconvenience of short gaps for 'occupation' on their passports and call themselves this) can also estimate the total time since the formation of the *r*-process elements by looking at how much of two kinds of uranium exists now. Uranium-235 decays on a time scale of 710 million years, whereas uranium-238 lasts much longer

(4.5 billion years). Both are formed from *r*-process elements in roughly equal amounts. Uranium-238 is 138 times more abundant than uranium-235. It can be calculated that approximately 6.5 billion years must have elapsed since both kinds of uranium were formed.

The gold you own as a wedding ring, watch or tooth filling is thus the seed of some supernova which occurred some 6.5 billion years ago, then drifted in space for up to half a billion years, condensed as part of a protosun but was rejected from the body of the Sun to become part of the solar nebula which then condensed into the planets (including the Earth) and which finally found itself by chance geological processes near enough to the Earth's surface to be extracted by gold miners.

Metal-rich, metal-poor

Although the vast majority of stars which we see now contain the different elements in relative proportions which closely match the proportions in the Sun, astronomers have found stars which are deficient in some of the total number of elements other than the hydrogen and helium produced at the origin of the Universe.

Thus, although, say, the ratio of the amount of iron to nickel is the same in most stars, the total amount of iron is highly variable from star to star.

Ignoring chemistry, astronomers have traditionally called all elements other than hydrogen and helium 'the metals', and stars with less than normal amounts of iron and so on are termed *metal-poor*. The lines of iron and other metals in their spectra are weaker than in other comparable stars, signifying that there is a smaller quantity of metals present to produce the spectral lines. Conversely *metal-rich* stars have stronger than normal metal lines.

From the scenario of element creation that has already been sketched, the reader can guess that the older stars are metal-poor, having formed early in the history of the Galaxy before there had been many supernovae to make metals, while the very youngest stars are metal-rich, having formed recently from interstellar material which has been enriched by the detritus of supernovae throughout the whole of galactic history. The Hyades star cluster, that prominent V-shape in the constellation Taurus, is an example of a metal-rich cluster of stars having twice the concentration of metals of the Sun and an age of just half a billion years (compared with ages in excess of 10 billion years for the oldest metal-poor stars).

There are, however, no stars found having no metals at all, so the amount of metals in the Galaxy appears to have increased steadily from a non-zero initial value. In fact, the number of metal-poor stars in general is embarrassingly low for simple scenarios.

Where did the initial metals come from if they were not created in the Big Bang? Where have the metal-poor stars gone? Or were so few formed that very few still exist?

One possible answer to these questions is that early in the formation of our Galaxy there may have been a wave of star formation which preferentially produced large numbers of massive stars. This idea is called Prompt Initial Enrichment (PIE), probably so that astronomers can make jokes about 'pie in the sky'. In one manifestation of the PIE theory, astronomers conjecture that there was a Population III of extremely old stars in addition to the Population II of old stars and Population I of new stars which both exist now. Population III stars might have been 100–1000 times the mass of the Sun. Since massive stars quickly turn into supernovae, there would have been a burst of metal formation,

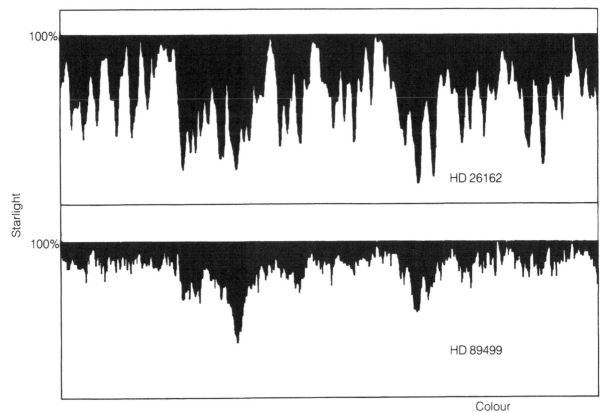

100%

Starlight

HD 26162

100%

HD 89499

Colour

FIG. 75. *Metal rich, metal poor. In these graphs of the spectra of two stars, the black areas (valleys) in each represent colours lost from the spectrum of the star because of absorption by metal atoms in its atmosphere. The deep valleys in each plot correlate from star to star, and show that, broadly, the same kind of atoms are present in each. But the strength of the valleys is vastly different in the two stars. The bottom spectrum is of* HD 89499, *a star with a deficiency of metals, and therefore showing relatively little loss of light due to absorption. By contrast, a heavy blanket of absorption lies across the top spectrum. This spectrum is of* HD 26162, *a star with a high metal content, similar to that of the Sun. Olin Eggen first drew attention to* HD 89499, *a star formed during the initial collapse of our Galaxy, perhaps 8 billion years ago. Mike Bessell estimates that it contains only 1% as much concentration of metals as the Sun. This diagram is based on AAT spectra made by Bob Fosbury and Mike Penston.*

and the wave of stars would soon have disappeared by becoming invisible black holes or faint neutron stars, pulsars whose ticking has long since faded to silence. This would imply that in the early years of the Galaxy there was a supernova every few days rather than every few decades as now.

With this answer it is difficult to maintain a balance between making this first generation of stars massive enough to do a convenient quick-disappearing act, but not making them so massive that they overproduce metals which would still be around now in embarrassing overabundance. Among the solutions proposed to cope with this problem is a guess that primordial material left over from the formation of the Universe has been, and presumably still is, falling into our Galaxy to dilute the metals in the interstellar medium. Little evidence for the existence of this in-falling gas is known. Indeed, if there were a large number of supernovae during, say, the first 2 billion years of the existence of the Galaxy, significant amounts of gas would be blown out of our Galaxy, rather than falling into it.

You, the supernova remnant

In this subject a large theoretical superstructure has been erected on a narrow keel of observational fact and it cannot be said that the so-called Ultimate Model of the chemical history of our Galaxy has yet been conceived, let alone detailed. (It is, apparently, called the Ultimate Model so that when asked what it is, astronomers can reply 'Um. . . .') According to Beatrice Tinsley

in a lecture in 1974 before her sad death from cancer in 1981, the scenario may proceed in five acts:

Act I. A Big Bang produces hydrogen and helium, in the early dense hot stages at the beginning of the Universe.

Act II. The Universe cools, becomes less dense and the lumps in it start to form galaxies. Actually, the lumps form galaxies more easily in the denser early stages than this later one, but nucleocosmochronologists turn a blind eye, a deaf ear and a cold shoulder to this difficulty.

Act III. Galaxies form into spirals or ellipticals. A portion of the gas in our Galaxy disappears during the first 2 billion years, forming new faint stars of low metal content, and the remainder collapses to a flat disc whose metal abundance is by then already half that of the solar system because of the first wave of supernovae.

Act IV. Our Galaxy continues to produce stars at a slowly decreasing rate for the next billion years (in the middle of which it produces the Sun). The rate of star formation drops slowly because gas is continually being locked up into neutron stars, white dwarfs and black holes, but the metal content of the gas remaining is continually increasing because each supernova enriches the gas with more metals.

Act V. The Sun having formed, a portion of the metals trapped in the rejected part of the solar nebula gives rise to the Earth, biological chemistry, evolution and the animal kingdom, including astronomers and readers of books. Looked at like this, people are the most interesting supernova remnants of all.

12

Cosmic rays

The crackle you hear and feel when you take off a sweater of synthetic fibre on a dry day is caused by static electricity. In the act of sliding the sweater over your head you rub its surface and fracture the atoms there, ionizing some and thus splitting off free electrons from the broken atoms. Some of the free electrons can be transferred from the sweater to you. You become charged with negative electricity, the sweater with corresponding positive electricity. If conditions are right, your hair stands on end. The spare electrons from the sweater are feeling one another's repulsion, pushing their fellows as far away as possible, separating your hairs as much as they are able and making them stand on end like a porcupine's quills, or Hamlet's hairs when hearing his father's ghost's tale of purgatory.

You may touch a metal object and the electrons on you will flee into the switch and into the Earth, scattering away from each other as far as they can. Suppose you stand still; suppose you are wearing rubber-soled shoes through which the electrons cannot pass. As you watch in the mirror you see your hair gradually settle into place as the excess of electrons in your body gradually leaves it. In a matter of minutes, they have dissipated.

How did the electrons on your body dissipate into the air? Dry air is a very good insulator, like rubber – electrons do not pass easily in air. What conducting threads reach through air from you to the Earth, along which electrons may pass?

Ionization of air

Attempts to understand this phenomenon started in 1900. The electron had been discovered three years earlier by J. J. Thomson and its role in the phenomenon of static electricity was being investigated by J. Elster and H. Geitel in Germany and C. T. R. Wilson in England. They established that even dry, pure air was not made up solely of complete atoms but that some of its atoms had broken into fragments consisting of electrons, and what was left, ions. Discharge of static electricity takes place by electrons skipping from ion to ion to reach the ground. But what causes the ionization of air?

The early experimenters attempted to find how the degree of ionization of air varied with atmospheric conditions, geographical location and time of day. For this purpose they developed a carefully insulated device known as an electroscope which would retain static electricity well except for the electrons that it discharged through the air. By watching its rate of discharge they hoped to obtain a clue to the amount of ionization of the surrounding air.

Suspecting that ionization of air was caused by radioactive rocks such as uranium-bearing ores, Wulf and A. Gockel took an electroscope to lakes and glaciers which, being of relatively pure water and ice, generated little radioactivity. The ionization fell considerably. Some ionization of air must therefore be caused by radioactivity. There was, however, a degree of ionization even over the thickest glacier, the deepest freshwater lake. Was there some residual radioactivity even in fresh water? Did moving the electroscope higher above the glaciers or lakes diminish the residual effect? Curiously, it did not.

If natural radioactivity from rocks was the cause of ionization, did this effect diminish with altitude? Wulf took his electroscope to the top of the Eiffel Tower in Paris (330 metres high). There was some decline but not enough. V. F. Hess in Vienna decided to investigate the effect of altitude by ascending in a balloon. His first flights to 1000 metres showed a small reduction of ionization in the air. By 1912 he had made ascents to more than 5 km and, to his astonishment, he found that the ionization actually began to *increase* at heights above 2 km. He concluded that radiation must be penetrating the atmosphere *from above*. The obvious source of such radiation was the Sun, but this had to be ruled out when Hess made an ascent during a partial eclipse. Although the Moon had obscured part of the Sun, there was no drop in the ionization of the air. Similarly, there was no difference between daytime and nighttime levels. Nor could the 23 hour 56 minute period of the Earth's spin beneath the stars be picked out – so whatever space radiation was ionizing the air could not come from any one object.

Explaining cosmic rays

These mysterious emanations were first termed *cosmic radiation* by the great American physicist R. Millikan, who began research in the subject after World War I. He also realized that cosmic rays were bringing an energy comparable to that of starlight to the Earth.

Gradually the complex properties of the cosmic rays became understood. The modern picture is that most cosmic rays incident on the Earth from space are atoms, stripped of most or all of their electrons and therefore are bare nuclei, moving at speeds close to that of light.

Cosmic rays are influenced by the magnetic forces in the Galaxy and the solar system. The lowest energy cosmic rays are deflected by magnetized clouds ejected from the Sun during periods of solar activity. The Sun's activity varies with an 11 year cycle. During its most active phase there are many large sunspots on its surface and many solar 'storms' which eject the clouds into the space between the planets. These clouds cocoon the Earth and shield it from the influence of interstellar cosmic rays, though the Sun itself may contribute some extra cosmic rays to the flux striking Earth. The magnetic field of the Earth

itself also plays a part in deflecting cosmic rays. The result is that the cosmic rays enter the Earth's atmosphere from random directions – they have lost all 'memory' of the direction from which they originated. It is as if cosmic ray astronomers had to observe through a frosted glass telescope which scrambled all the lights which entered it.

Those cosmic rays which do reach Earth's upper atmosphere collide with air atoms to make showers of electrons and other particles of matter which cascade down to the Earth's surface. These secondary particles are the ones which ionize air near sea level and cause the gradual discharge of static electricity from electroscope or tousled hair.

Much of the early study was of the secondary cosmic rays; only as balloons and rockets reached far above Earth did the properties of the primary cosmic rays become clearer. Present-day studies have used cosmic ray detectors in space, both on artificial satellites and on the Earth's natural satellite, the Moon. The latter experiments were set up on the Moon's surface during the Apollo 16 lunar flight and brought back to Earth afterwards.

The abundances of the different nuclei found among the cosmic rays follow closely the abundances of the nuclei determined from the cosmochemistry of meteorites and the like as discussed in the last chapter – hydrogen is the most abundant, and the 90-odd remaining heavier elements are less and less common as they increase in complexity. There are three major exceptions – the elements just heavier than helium, namely lithium, beryllium and boron, are a million times more abundant in the cosmic rays than in other matter in the Universe.

We saw in the last chapter that the creation of elements in the Big Bang could not get to lithium. Moreover, it turns out that nuclei of all three of these elements are readily destroyed at

temperatures of above a million degrees –
relatively low compared with those found in stars.
Far from lithium, beryllium and boron being
created in stars, therefore, any interstellar material
condensing into a star and then being returned to
space in a supernova or whatever has been purged
of these elements, unless they have been made on
the relatively cooler stellar surface or in space
itself. If these three elements can be made neither
in the Big Bang nor in stars, where can their
origin lie? The B²FH paper of the Burbidges,
Fowler and Hoyle named the process which
created these elements the *x*-process because it
was unknown.

The *x*-process may in fact be the travel of
cosmic rays through the interstellar material. The
cosmic rays contain carbon, nitrogen and oxygen
nuclei in relatively high abundance. As such nuclei
collide with hydrogen and helium in interstellar
space, the individual protons and neutrons are
rearranged into several fragments of different sizes
including lithium, beryllium and boron nuclei.
(This process is also called *spallation*.) The cosmic
rays might therefore be expected to contain an
excess of spallation products. They contain not
only the great excess of lithium, beryllium and
boron, but also other spallation elements such as
chlorine and manganese.

From the amount of these elements in cosmic
rays, cosmic ray physicists can establish through
how much interstellar matter the average cosmic
ray has travelled in order to produce the observed
quantity of spallation products. The average
cosmic ray has been travelling in the Galaxy for
some 10–100 million years. After this time it
leaks from the Galaxy into intergalactic space.
Though a long time on a human time scale,
10–100 million years is short when compared to
the age of the Universe (measured in units of
10 billion years) and the lifetime of most stars. If

it is to the stars that we must look for the
creation of cosmic rays, it is to the short lived,
massive ones that we must direct our attention,
and these are the ones which turn into
supernovae. In fact Baade and Zwicky calculated
that supernovae occur often enough and have
enough energy to account for the supply of all the
cosmic rays in the Galaxy. A modern calculation
indicates that only 3% of the energy of
supernovae need be used to make cosmic rays.

The way in which they do so is not clear.
Some believe that cosmic rays are injected into
space during the supernova explosion itself.
Others say that it is not the supernova which
supplies cosmic rays to the Galaxy but the
rotating neutron star or pulsar, which the
supernova forms. Perhaps a mixture of these
processes is responsible for pumping cosmic rays
into space to stream about the Galaxy, and,
before they escape from it, possibly to collide with
Earth.

Gamma rays

When cosmic rays and relativistic electrons
interact with other material or with other forms
of energy, such as magnetic fields, they can
produce very energetic photons called *gamma
rays*. Unlike cosmic ray particles, gamma rays
travel in straight lines from their point of origin.

Gamma rays are absorbed by the Earth's
atmosphere and do not reach the Earth's surface.
The more energetic gamma rays which penetrate
the outer layers of the atmosphere can be studied
with gamma ray telescopes suspended from high-
flying balloons. The most energetic gamma rays
produce effects in the atmosphere which percolate
to the ground, and high energy gamma rays can
be studied there. There are two effects. Gamma
rays produce fast-moving electrons as they hit the
upper atmosphere. These electrons move at speeds

which are (of course) less than the speed of light in a vacuum, but are actually more than the speed of light in air, since light travels slower in transparent material than in free space. The electrons produce bursts of light called *Čerenkov radiation*, and this can be detected as brief flashes from the night sky. If the original gamma rays are energetic enough, the electrons which they produce in the upper atmosphere form a shower which reaches the ground. These so-called *extensive air showers* of electrons (which occur at a given place at a rate of a few per hour) can be detected in arrays of scintillation counters laid out on the ground over a few hundred metres. The time delays as the extensive air shower sweeps across the array give the direction from which the shower came. For instance, if the shower is detected in the easternmost counters before the western ones, then the gamma ray source is rising in the east. The time delays are very small, at most one millionth of a second, but can be measured accurately enough to tell the direction of the source to about 0.1 degree.

The most complete results, however, have come from space studies, particularly from a satellite called COS-B. This satellite saw gamma rays coming most extensively from the Milky Way, the galactic plane region. They are produced by cosmic ray interactions with the hydrogen gas which concentrates in this region.

COS-B also picked out two dozen gamma ray sources – bright spots on the gamma ray map of the sky. The ability of a gamma ray telescope to pinpoint an object is poor and so the image is blurry. The result is that the region of the sky from which the gamma rays come is not very precisely defined, and it is therefore difficult to be sure of their origin; the identity of most gamma ray sources remains a mystery. For instance, the strong source known as Geminga is still

unidentified despite detection by COS-B, SAS-2 and even by ground-based studies. Indeed, the word *geminga* is from the Milanese dialect and means *nothing there*. (The region in which Geminga lies – its *error box* – contains a weak X-ray source, which has been suggested as the identification, and the X-ray source can be seen optically as a faint star; but its association with Geminga may be just a fortuitous line-of-sight coincidence.)

A quasar, 3C273, is one gamma ray source which has been identified but most are inside our Galaxy. The Crab and Vela pulsars are two examples. These can be identified by the effect of their pulses on the gamma rays from their direction – the sources pulsate with periods of 33 and 89 milliseconds respectively. Because these pulse periods are unique, they are referred to as time-signatures, and they identify the gamma ray sources unambiguously. An astonishing discovery in 1983 was the identification of a gamma ray source by means of a 4.8 hour time-signature.

The 4.8/h time-signature of the gamma rays, which were detected to come from the constellation of Cygnus, can immediately be associated with an X-ray binary star called Cygnus X-3. Gamma rays from the stars were first detected by Soviet astronomers during the balloon flight of a gamma ray telescope in 1972. In fact, the team was able to provide a small correction to the period of the star which was later found to be correct. Sadly, the Soviet results were not widely believed in the West, until confirmed in 1984 by experiments at the Universities of Kiel (Germany) and of Durham (England). Those experiemnts detected higher-energy gamma rays by the extensive air-shower method. Although Cygnus X-3 is radiating between 100 000 and 100 000 solar luminosites of gamma radiation and a few systems like it can supply the whole of the gamma rays seen in our Galaxy.

The Cygnus X-3 source lies 30 000 light years away behind a mass of interstellar material which hides its optical counterpart. However, infrared radiation and radio waves penetrate the dust, and Cygnus X-3 has been seen at these wavelengths. It is a binary star, containing an accreting neutron star, orbiting another star with the period of 4.8 hours. In one theory for the origin of the gamma rays, the neutron star is a young magnetic pulsar spinning several hundred times per second. It is supposed to emit ultra high energy particles which interact with the atmospere of the companion star to make the gamma rays. In 1973, Cygnus X-3 underwent a radio and X-ray outburst; it was among the brightest radio objects in the sky. It evidently went through an unusual nova-like stage, but the details have never been satisfactorily explained. One speculation is that this was the supernova-like event which created the young pulsar whose effects we now see.

The influence of cosmic rays

Do the cosmic rays have more profound effects on the Earth than influencing the discharge of static electricity? In fact it appears that they do. Because of supernovae, archaeologists have a time scale for dating in prehistory.

The spallation reactions occurring as the cosmic rays collide with Earth's atmosphere produce neutrons, most of which are eventually absorbed by the predominant gas in the atmosphere, nitrogen. In this way the nitrogen turns into a carbon isotope, of a kind called carbon-14 (having an excess of two neutrons over the more usual carbon-12). Carbon-14 is unstable and decays to nitrogen-14, half of it changing every 5568 years. Though the carbon-14 is produced high in the atmosphere, at an altitude of

10–15 km, it combines with oxygen to make carbon dioxide, and diffuses rapidly to Earth's surface.

Carbon dioxide is a gas which is abundant in the atmosphere and which is 'breathed in' by plants to take part in their cell-building processes. In this way, therefore, a small proportion of the carbon atoms which the plants accumulate during their lifetimes consist of radioactive carbon-14 atoms. Because plants are eaten by animals, all living organic material contains this proportion of radioactive carbon. When the plant or animal dies, the assimilation of carbon stops and the carbon-14 begins its long, steady decay to ordinary carbon. If the organism has recently died, its cells will contain their full abundance of carbon-14; if it has been dead a long time, the proportion of carbon-14 will be small. The ratio of ordinary carbon to carbon-14 is therefore a clock which runs down at a known rate.

In the cases where parts of the dead organic matter remain preserved – as timbers in a house, or bones in a midden, for example – this clock can be used to measure with some accuracy the year in which the tree was felled or the animal died. This technique, known as radiocarbon dating, can be used to estimate the age of material up to 20 000 years old, and is therefore of inestimable value in archaeology.

This method relies on the level of cosmic radiation having been constant for many thousands of years. Suppose there is a sudden significant burst of cosmic radiation from space, producing more carbon-14 than normal. When this carbon-14 is assimilated by living organisms, their radioactivity level is suddenly much higher. After death, the organic material still exhibits a proportionally higher radioactivity level. It would seem to the archaeologist that the clock was still tightly wound, and that the organism was more recently dead than was the case.

There is good evidence that the production of carbon-14 has indeed been variable in the past. Evidence for the variation comes from using two different methods to estimate the age of wood from the incredibly long-lived California bristlecone pine.

One method is radiocarbon dating, while the other is called dendrochronology – counting tree rings. Each year, a tree adds a new ring to its circumference, and the thickness of the ring represents the growth of the tree that year – a thick ring in good years, a thin one in bad. Counting these rings offers a potentially accurate way of determining the age of a tree whose date of death or felling is known, and bristlecone pines as old as 4600 years have been found. By overlapping ring patterns from many bristlecone pines, representing the alternation of good and bad years for the growth of trees in California, C. W. Ferguson has established the age of wood which formed 8000 years ago. Radiocarbon dating of wood as old as that, however, seems to suggest that it is some 900 years younger. Only over the last 2500 years do the two dating methods agree, on average. Evidently the rate of production of carbon-14 was once higher than it is now.

Was this the result of a sudden burst of supernova activity increasing the number of cosmic rays in the Galaxy? The Czech geophysicist V. Bucha offers another explanation: that it was due to changes in the Earth's magnetism and its corresponding ability to shield the atmosphere from cosmic rays.

There is no real evidence for a significant change in the cosmic ray flux in historical times. The record of 'fossil' cosmic rays has been extended back to 10 million years by a count of the number of tracks left by cosmic rays as they struck meteorites orbiting the solar system, showing no major changes in that time interval. Before then, the cosmic ray intensity in the Galaxy could have been different from now, and presumably, was much larger very long ago during the first wave of formation of massive stars in the Galaxy and the consequent high rate of supernovae. We have no direct evidence about the cosmic ray rate at this time.

Did a supernova kill the dinosaurs?

In 1957 Iosif Shklovsky considered what would occur if a supernova exploded near to the Sun, within say 10 parsecs. The supernova would shine at magnitude −20 for a matter of months (not as bright as the Sun but far brighter than the Full Moon). After a few thousand years, the gaseous envelope of the supernova, ejected from it, would pass across the solar system. If the supernova were like the Crab Nebula, the synchrotron radiation trapped within the filaments would be seen as bright as the Milky Way, filling half the night sky. There would be no dynamical changes in the solar system – the amount of mass striking the planets would be too small to deflect them from their orbits at all. However, the solar system would be within the supernova shell for some 10 000 years, and the density of cosmic rays would be 10–100 times its current value. There would be much more cosmic radiation striking the Earth's atmosphere and much more radiation at sea level caused by the secondary cosmic rays.

Shklovsky has pointed out that such an increased level of background radioactivity might have played a part in the evolution of life on Earth. Evolution proceeds as a result of chance mutations in individual biological organisms. Some mutations better fit the organism to survive and, during its longer lifetime, it has a better chance to perpetuate the mutation to future generations. Most mutations, however, are unfavourable, particularly when the change to the genetic material has been gross. It might be expected that a large increase in radioactivity would produce many gross changes in genes and thus tend to cause organisms to die and become extinct.

Two-thirds of the mean radioactivity on Earth is caused by terrestrial factors, mostly natural radioactivity in rocks at the Earth's surface. One third is caused by cosmic rays. If the cosmic ray intensity increased 100-fold, the background radioactivity would be some 30 times increased. Shklovsky speculates that the dying out of the giant reptiles at the end of the Cretaceous period was the consequence of such an increase caused by the sudden bathing of the Earth in the ejecta of a close supernova unwitnessed save by the uncomprehending eyes of doomed prehistoric dinosaurs. This increase of radiation corresponds to a radiation dose of about 1 roentgen per year. The lethal dose for most present-day laboratory animals is of the order 500 roentgens, so the radiation might or might not kill individual animals, depending on their exposure, lifetime, size, and susceptibility to radiation; it would certainly affect their health and increase the proportion of abnormal births.

What evidence is there for such a change in the biosphere? Geologists and palaeontologists refer to the discontinuity in the history of the Earth's crust 60 million years ago as the Cretaceous–Tertiary boundary. Within a few hundred thousand years, the reptiles lost supremacy in the evolutionary battle and gave way to the mammals, culminating (as human beings put it) in the primates, including humans; there were other correlated changes in the mixture of a wide variety of species. The abruptness of the discontinuity is a cause of much argument, and

provokes an emotional division into two attitudes. One, under the name of *gradualism*, holds that all changes in Earth's history were continuous, gradual and evolutionary: this attitude is in antithesis to *catastrophism* which sees the changes as abrupt and dramatic, with periods of stability between the changes. The attitudes were developed during the time in geological study when it was natural for scientists (like Baron Georges Cuvier, with whose name catastrophism is linked) to relate natural phenomena to biblical versions of Earth's history, and The Flood is one 'catastrophe' whose results were repeatedly 'identified' in the geological record. The gradualists, typified by C. Darwin and A. Wallace, won the historic argument, but nevertheless astronomers have speculated about several possible catastrophic events which would affect Earth's history. The impact of a large meteor or asteroid on the Earth is one such possibility.

From spacecraft sent to photograph other planets (Mercury, Mars) and satellites (the Moon, the satellites of Mars, Jupiter and Saturn) it is known that impacts by meteors have dramatically affected at least the early history of the surface of the planets. Craters from meteor impacts are harder to find on the Earth than on the airless surfaces of other planets, because the surface of the Earth is subject to steady erosion and rebuilding: these processes obliterate craters. However, about 100 meteor craters have been identified on Earth, the largest of them about 100 km across, in rocks from 20–200 million years old. An asteroid 10 km in diameter, producing a yet-to-be-identified crater about 175 km in diameter (maybe the asteroid fell in what is now sea), would eject enough dust into the atmosphere to cut out most of the light normally reaching Earth's surface. This would reduce the quantity of photosynthesis in plants, causing the food chain to break at its beginning and precipitating extinctions of whole species by hunger.

A large solar flare is another possible astronomical catastrophe, which would have effects upon the atmosphere, the ozone layer and the radiation balance, and which therefore could trigger evolutionary disasters. Shklovsky's speculation about a nearby supernova suggests another possible astronomical catastrophe.

An effect on genetic material by the increase of cosmic radiation is only one consequence of a nearby supernova. The Earth's ozone layer would be depleted and large amounts of nitric oxides would be formed in the lower atmosphere. Visible radiation from the Sun would find it more difficult to reach the Earth's surface. Photosynthesis would become more difficult for plants, the temperature of the Earth's surface would drop, the water of the oceans and lakes would become locked up in the ice caps and there could be a worldwide drought.

The magnitude of this catastrophe depends on how near the Earth lay to the supernova when it exploded. The Earth passes through the spiral arms of our Galaxy, in which there is an abundance of massive stars of the sort that become supernovae, every 50–100 million years. The passage takes about 10 million years. There is an evens chance that during this time a star will explode within 10 parsecs. Similar events would have occurred about 100 times in the history of the Earth.

There is no general agreement by palaeontologists that an astronomical catastrophe caused the Cretaceous–Tertiary boundary. Other theories suppose that there was a coincidence of terrestrial events – volcanoes, marine regressions, climatological changes. If the specific explanation of the Cretaceous–Tertiary boundary is unclear, nevertheless the repeated passage of the Earth through new and active supernova remnants certainly had effects which are part of our Earth's and our species' history.

In an interesting development of the theory that cosmic rays cause mutations, George Michanowsky has proposed that cosmic rays from the Vela supernova, occurring at about 8000 BC, triggered off a new awareness in men's minds and precipitated the technological era which we are still witnessing. If this is right, the Vela supernova would be like the intelligence-donating obelisk in Arthur C. Clarke's book and movie, *2001 – A Space Odyssey*, and might have had a physical effect on mankind comparable to the mental liberation sparked by Tycho's supernova in 1572. The credibility of Michanowsky's theory is undermined by the distance of the Vela supernova remnant from Earth – more than 50 times the distance of the hypothetical supernovae about which we have just been speculating. The Vela supernova's effects on Earth were therefore 2500 times less than these. The cosmic radiation which it produced is therefore probably less than the background radiation from other sources and could hardly have had a noticeable effect.

Nonetheless it must be true that cosmic rays, by working on the genes of all species, inevitably cause mutations. Since, in fact, the natural radioactivity in rocks is a consequence of the formation of radioactive heavy elements in supernova explosions, and since the cosmic rays are generated in supernova explosions, it could be said that, having played a crucial part in producing and distributing the chemical elements which make life possible at all, the supernovae are responsible for life's further evolution. Since the Sun now lies on the edge of a spiral arm of the Galaxy and, a majority of astronomers believe, is about to enter it, there is a 50–50 chance that a significant change in the life forms now inhabiting the Earth will take place during the next 10 million years.

13

Black holes from supernovae

The bigger they come, the harder they fall.
R. FITZSIMMONS, CHAMPION BOXER

So far in this volume, we have given evidence that many supernovae produce neutron stars. Some are seen as pulsars like those of the Crab Nebula, the Vela supernova remnant, LMC 0540–69.3 and MSH15-52. Some are seen as point-like X-ray sources: 3C58 and RCW103 are two supernova remnants containing these. In one case, CTB109, a supernova remnant is associated with an X-ray binary star with a neutron star component. Some supernova remnants are filled with synchrotron emission from an active pulsar. High-speed pulsars can plausibly be connected with the disruption of binary stars in which one star explodes as a Type II supernova. The general statistics of pulsars are compatible with them being formed from supernovae: the 330 known pulsars generally lie in the galactic plane (like Type II supernovae and the massive stars from which they come). The birth rate of radio pulsars, according to a survey by Jodrell Bank radio astronomer Andrew Lyne, is that one is formed every 20–50 years in our Galaxy, at about the same rate that supernovae occur (whether of Type I or Type II or both combined).

Do *all* supernovae produce neutron stars? How do we account for the failure of diligent searches for the stellar remnants of supernovae in almost every extended supernova remnant? We fail to see most pulsars because their beams never sweep across Earth. Pulsars have large slingshot velocities and move rapidly away from the centre of the remnant. Maybe Type I supernovae (about half the total) produce relatively inactive white dwarfs, or even no stellar remnant at all. All these effects reduce the numbers of possible detections of neutron stars in supernova remnants.

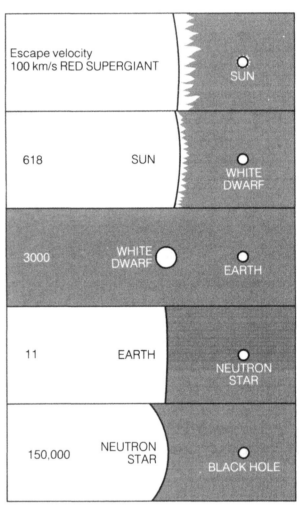

FIG. 76. *Escape velocity. Escaping from the Earth needs an 11 km/s impulse and is relatively easy because Earth's mass is small. A red supergiant is the easiest star to escape from because its size is large. Escaping from other stars becomes more and more difficult as the star becomes more and more compact. Escaping from the most compact objects known, black holes, is impossible, because it requires a speed greater than that of light.*

But the definitive problem seems to be Cas A. Shklovsky pointed out that the evidence from the oxygen abundance in the fast-moving knots implies that Cas A was formed from a Type II supernova explosion of a massive star, and there has not been time for any pulsar formed by the explosion to move from the centre. Cas A contains no radio pulsar, no X-ray binary, no X-ray point source and no synchrotron emission from an active pulsar. Perhaps some Type II supernovae produce other, stranger stars which we cannot see in these ways.

How to escape

A stone tossed in the air loses speed and momentarily halts at its highest point before plummeting back to Earth again. When hurled with greater force, the stone rises higher but is still drawn back by the force of gravity. There is clearly a relation between the initial speed given to the stone and the height to which it rises: scientifically speaking, that momentary halt at maximum height is where the stone's potential energy due to the force of gravity exactly equals the kinetic energy it was given as it was thrown. The question arises: what gravitational energy does the stone have at higher points above the Earth and what speed is it necessary to give the stone to throw it arbitrarily far above the Earth, to 'infinity'? Is it possible for the stone to be hurled with sufficient speed for it to leave the Earth completely and escape its gravitational pull? It is, and this speed is called the *velocity of escape*. The velocity of escape from the Earth's surface is 25 000 miles per hour: an object thrown with this speed from Earth does not fall back.

How does the velocity of escape vary from place to place – from planet to planet, star to star? The larger an astronomical body, the less the force of gravity at its surface; but the more massive it is, the greater its gravitational force. Put in mathematical terms, the velocity of escape is proportional to the square root of the mass of the body divided by its radius. Jupiter has 318 times the mass of the Earth and a radius 11 times larger than the Earth's: the escape velocity from Jupiter is therefore the square root of 318/11, which is 5.4 times that from Earth.

The escape velocity from the Sun is more than 1 million miles per hour: though larger than the Earth it is much more massive and so has a larger velocity of escape. Curiously, although other dwarf stars like the Sun range in mass from a few hundredths of the Sun's mass to 60 times its mass, the range of sizes almost exactly compensates for this, with the result that the escape velocity from all other dwarf stars is not much different from that of the Sun.

Only when stars have finished the phase in which they burn hydrogen and expand to become red giants and supergiants does their escape velocity alter much from the solar value. In a red supergiant the escape velocity can be less than one sixth the solar value. This is the reason why matter can relatively easily escape from stars at the red giant and supergiant stage so that some massive stars can bring their mass under the Chandrasekhar Limit and become white dwarfs in their subsequent evolution.

As you might expect, white dwarfs have high escape velocities. Even though their masses are similar to that of the Sun, they are denser and more compact, with radiuses typically one hundredth that of the Sun. From the square-root law, therefore, the escape velocity is typically 10 times as high as the Sun's. If matter is to escape from a white dwarf, such as in a nova explosion, when the white dwarf's outer envelope is thrown off, the explosion must be energetic enough to give the envelope an escape velocity of

Table 8. *The velocity of escape*

		Velocity of escape* (km/s)
Earth		11
Moon		2.4
Mars		5
Jupiter		58
Sun		618
Typical stars	Blue dwarf	900
	Red dwarf	400
	Blue supergiant	900
	Red supergiant	100
Collapsed stars	White dwarf	3000
	Neutron star	150 000

* 1 km/s = 2240 miles/hr; speed of light = 300 000 km/s.

thousands of kilometres per second.

The concept of escape velocity is useful in describing not only the energy required to blast matter off the surface of a star, but also the loss of energy by radiation as it is emitted from the star. Radiation too loses energy as it travels against the force of gravity. This fact is not obvious from ordinary experience but is a feature of Einstein's General Theory of Relativity. Although very small in the Earth's gravitational field, the loss of energy has nonetheless been measured in gamma rays travelling up a mine shaft.

The lower the energy of radiation, the longer its wavelength. As light loses energy climbing out of a gravitational field, therefore, its colour shifts towards the red end of the spectrum. This phenomenon is termed the *gravitational red-shift*. The fraction of energy lost by radiation as it leaves the surface of a star is the square of the velocity of escape measured as a fraction of the speed of light. The gravitational red-shift is not measurable in most stars, as we can see by taking the Sun as an example. Since the velocity of escape from the Sun is about 600 km/s and the speed of light is 300 000 km/s, the fraction of energy lost by light as it leaves the Sun's surface is only $(600/300\,000)^2$ or one part in a quarter million.

Although the Sun's gravitational red-shift is barely detectable, soon after white dwarfs had been discovered, Arthur Stanley Eddington in 1924 pointed out that they had a large escape velocity and that measurable red-shifts could be expected in their spectra. The escape velocity from a white dwarf is typically 3000 km/s (Table 8), so the red-shift is $(3000/300\,000)^2$, or one part in 10 000. Thus H-alpha light emitted from hydrogen at a dense white dwarf's surface with wavelength 6563 angstrom units might be seen by an observer on Earth with wavelength almost 6564 angstroms, a small but measurable increase. There was, however, a confusing detail: there is no way of distinguishing a gravitational red-shift from the more familiar Doppler red-shift caused by possible motion of the white dwarf away from Earth, at a speed of a few tens of kilometres per second or so, unless the speed could be accurately known, and allowed for.

The only way round this problem was by finding white dwarfs which were part of binary systems, sharing a common motion with other ordinary stars. W. S. Adams measured the shift in the spectrum of the white dwarf Sirius B, which is in orbit around the bright star Sirius itself. The speed of Sirius is 8 km/s towards the Sun; the speed of motion of Sirius B around Sirius could be accurately calculated from knowledge of its orbit, and the red-shift still unaccounted for was 21 km/s compared with the value of 20 km/s calculated by Eddington. The red-shift of 40 Eridani B, a white dwarf in orbit around the star 40 Eridani, has been similarly measured by D. M. Popper, with good agreement with the theoretical value.

Measuring the gravitational red-shift from the surface of a neutron star such as the Crab Nebula pulsar would be very interesting, since the wavelength of light emitted from its surface would be much changed from its original value. Formerly invisible ultraviolet light would be red-shifted into the visible part of the spectrum, and the shift would give astronomers a way to estimate the mass and radius of a pulsar. In fact there are no atoms at the surface of a neutron star to emit light of a distinct wavelength: the red-shift cannot be measured, although it would be very large.

Possible enormous gravitational red-shifts may be responsible for the behaviour of high red-shift quasars, mysterious star-like objects lying among other galaxies. Some, perhaps most, astronomers argue however that the red-shift of quasars is caused by their motion in the Universe as they participate in the explosion of the Big Bang and recede from us, and that the high values of the red-shifts of quasars are a consequence of their great distance.

There is a limit to how large a gravitational red-shift can be. In climbing the gravitational field of a star the radiation cannot lose more energy than it possesses. The ultimate red-shift occurs when the fraction of energy lost by radiation is 100% which occurs when the escape velocity at the star's surface is equal to the speed of light. It is possible, then, to conceive of a star with such a powerful gravitational field that radiation cannot leave the star. Nor can matter leave the star since to do so it would have to travel at the velocity of light, and according to Einstein's Theory of Relativity nothing material can travel at the speed of light or faster.

Nothing at all could ever leave such a star because its gravity would be simply too strong – it would be black because no light could leave it, and it would be a hole because anything dropped in could not get out. Hence the name of such a star – the *black hole*.

How to make a black hole

How would such a star be formed? Take an ordinary star (in what is called a 'thought experiment', one which is impossible actually to carry out except in imagination) and compress it in a vice in an attempt to make it smaller and thus increase its escape velocity and its gravitational red-shift. The star will resist this attempt by increasing its internal pressure – reducing the star's size causes its atoms to pound faster and more often on the jaws of the vice to attempt to pry it apart. The star will remain in equilibrium in a kind of war between gravitational self-attraction and its repelling internal pressure forces.

But suppose that the gravitational vice clamped tighter and tighter, squeezing the interior of the star smaller and smaller. A neutron star might be the result. It too would resist further compression but less enthusiastically than ordinary stars: the repelling pressure mechanism is not so easily able to respond to any increased self-attraction. In fact the more massive neutron stars are less able to increase their internal pressure in response to further gravitational contraction. Beyond a certain critical mass in fact, they cannot respond at all: their internal pressure is at a maximum and cannot be increased. The gravitational self-attraction of such a star is always larger than the repelling pressure forces. The star cannot support itself and simply shrinks smaller and smaller in a continuing collapse. As it does so the velocity of escape at its surface increases until it reaches and passes the speed of light, at which stage whatever is inside the surface is shut off from the rest of the Universe forever. The star has become a black hole.

Since, in a sense, a black hole can be formed as a kind of extreme neutron star, it is natural to look to the same process that forms neutron stars to form black holes. Black holes appear to be formed in supernova explosions in the case where the mass of the stellar core which begins the collapse exceeds the critical mass of neutron stars, a few solar masses.

Are black holes the missing mass?

Is there evidence that black holes exist? There is certainly evidence for the existence of invisible matter in our Galaxy. Its effect can be seen in the motions of stars. As it orbits the Galaxy, a star is subject to the gravitational pull of all the others in the Galaxy. The stars which we see away from the Milky Way are above or below the Sun, which lies close to the central plane of the Galaxy wherein lie most stars. Stars above or below the galactic plane are pulled back towards it and the force which pulls them can be estimated from the speeds of stars at different heights above the plane. The higher stars generally move more slowly, just like stones flung from the surface of the Earth. If the force pulling these stars back is known, we can calculate the amount of mass in the galactic plane required to produce such a force.

The answer is that in every cube of space of 10 light years on the side near the galactic plane in the vicinity of the Sun there are on average 4.5 solar masses of material. About 1.8 solar masses of that can be accounted for as visible stars. A further 0.9 solar masses can be detected as interstellar gas in the form of hydrogen atoms. Approximately 2.7 solar masses is invisible. Part of this mass is certainly hydrogen molecules which emit no identifiable radiation received on Earth; part is undoubtedly neutron stars which have stopped being pulsars and part is former white dwarfs which have cooled to invisible black dwarfs. There has been speculation that some fraction at least of this so called 'missing mass' in

the Galaxy is in the form of black holes. Potentially, then, there may be large numbers of black holes waiting to be found.

On the other hand, R. Ruffini and J. A. Wheeler wrote in 1971,

of all objects that one can conceive to be traveling through empty space, few offer poorer prospects of detection than a solitary black hole . . .

By its very nature, the black hole cloaks itself with invisibility. No evidence of its existence beyond its surface can ever be seen, no action on its surface can send a message to proclaim its occurrence. The message carrier – radio pulse, light flash, cosmic ray or what you will – cannot escape the black hole's pull of gravity, which is why the surface of a black hole is called the *event horizon*. Occurrences within it are never seen.

If we cannot see a black hole itself can we see its surroundings? What happens when something encounters a black hole? What would happen if a black hole were to draw material into itself by gravitational force?

How the invisible shines

In an effort to describe how a black hole could be found, Iosif Shklovsky in 1967 considered what would happen if a star were to orbit a black hole. The black hole and its companion would circle each other with little effect on each other at first, beyond their mutual orbiting. A distant astronomer might wonder why the radial velocity of the companion star changed periodically and deduce that the star was orbiting another which he could not see. (Such stars are known as single-line spectroscopic binaries and it is possible that some of their invisible companions are black holes, though no doubt the vast majority of invisible companions are simply fainter stars outshone by the star which can be seen.) But there will come a time when the ordinary star circling the black hole will begin to turn into a red giant or supergiant. Its atmosphere will leak onto the black hole. Just as it does when a star's atmosphere falls on to a neutron star, the compressed gas will be heated to temperatures of millions of degrees and will radiate X-rays. To find a black hole, said Shklovsky, look among the X-ray stars. But how can we distinguish X-ray emissions from a black hole from those from a neutron star?

The major fact which would distinguish the black hole would be its mass, perturbing the companion star by the force of gravity. If the mass of such an X-ray source, measured by the size of the perturbation of the companion star, was larger than the extreme upper limit to the allowable masses of a neutron star, then the X-ray source might be a black hole.

The first such X-ray source known, the invisible companion to a large but ordinary star, is Cygnus x-1. First observed by rocket- and balloon-borne X-ray telescopes in the mid-1960s, Cygnus x-1 was one of the first X-ray stars studied by the Uhuru satellite in 1971. With this satellite a group of X-ray astronomers, led by R. Giacconi and H. Gursky, located Cygnus x-1 in a small area of the sky in which radio astronomers (L. Braes and G. Wiley in Holland and R. M. Hjellming and C. Wade in the US) had spotted a radio star which had not been visible before. At the same time that the radio star turned on, Cygnus x-1 changed its X-ray character, proving that the X-ray star and radio star were one and the same object.

At the same position of the radio star was a visible star, numbered 226868 in the Henry Draper catalogue (HD). HD 226868 was immediately found to be a hot supergiant star,

about 30 solar masses, but not peculiar in any way. The blue supergiant was so normal that two groups of astronomers who studied HD 226868 concluded that it was also a red herring, and nothing to do with Cygnus x-1. But two other groups of astronomers in England and Canada kept observing the star to see whether it changed at all. Astronomers at the University of Toronto and at the Royal Greenwich Observatory simultaneously published their findings that it did. The star had a cycle of radial velocity change which lasted 5.6 days as it orbited an invisible star, alternately approaching and receding from Earth so that the lines in its spectrum were blue- and red-shifted by the Doppler effect.

It was not possible immediately to say precisely what the mass of the invisible companion was, since the inclination of the orbit of the binary star was unknown and the astronomers could not tell whether they saw the full motion (orbit seen edge-on) or a small part (orbit seen nearly face on), but clearly the mass of the invisible companion had to be at least 6 times that of the Sun to swing HD 226868 as it did. This minimum is more than is possible if the invisible companion were a neutron star, being more than the critical mass of neutron stars, namely 3.2 solar masses.

Thus Cygnus x-1 fits in detail the scenario for the discovery of a black hole in a binary star system, outlined by I. Shklovsky in 1967 as a solution before the existence of the problem.

This is not to say that Cygnus x-1 *must* be a black hole. Perhaps the X-ray emitter is a neutron star (of 1 solar mass) orbiting a dim normal star (of, say, 5 solar masses) which itself orbits the bright and visible supergiant. No trace of the dim star has been found, and the black hole explanation accounts for the observed facts. It has the attraction that it was proposed before the facts were known; it does not suffer from the suspicion that it has been shaped to fit in. Perhaps a black hole, exotic though it may be, has indeed been discovered lurking in Cygnus x-1, formed by a long-past supernova and enabled to shine by devouring its companion.

The astonishing happening in 1408

It would of course be very exciting to identify the supernova which formed the black hole in Cygnus x-1. A number of Chinese and Japanese records suggesting a supernova in Cygnus in 1408 may form that very identification.

The Japanese record comes from the diary of the Honourable Noritoki, in which he noted the daily events of his life between 1405 and 1410. On 1408 July 14 (which he called Ohei Year 15,

FIG. 77. *Cygnus x-1. Gas from a large blue supergiant star,* HD 226868 *is overflowing on to a companion black hole. Its atmosphere leaks towards the black hole, encircling it before falling in and being heated to emit X-rays.*

Month 6, Day 21, Cyclic Day Tsuchinoe-Inu), Noritoki noted that the weather was fine and that he and his son had received two visitors. His son, Kurabe, had visited the temple, he wrote, and had

learned about the appearance of a guest-star etc. Eight evil features were listed and reported to the authorities by the astronomical doctors. It is an astonishing happening.

What the eight evil features were, or what lies behind the tantalizing '*et cetera*' in the diary entry no one can now say. But the record establishes the first sighting of the guest star of 1408, which was seen in total for at least 102 days, as judged from two further records from different parts of China. These give some details about the position of the star. The *Ming Shih* (*History of the Ming Dynasty*) says:

Reign Yong-le, year 6, month 10, day Geng-chan [1408 October 24]
To the south east of the asterism Niandao there appears a star as big as a lamp, yellow in colour and smooth in lustre. It does not move.

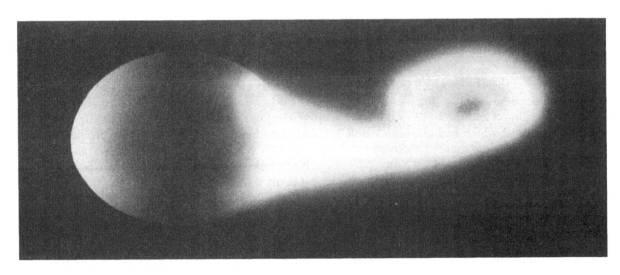

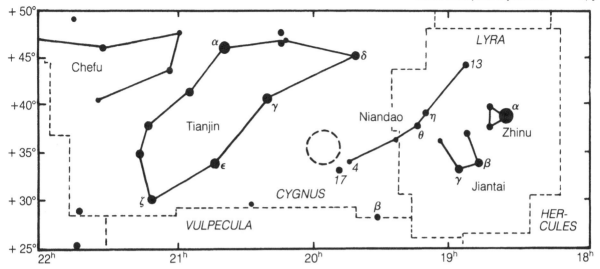

FIG. 78. *Supernova of 1408. 'To the southeast of Niandao there appears a star as big as a lamp . . .' is the statement in the Ming Shih, referring to a guest star of 1408 Oct 24. Niando is an asterism east of Alpha Lyrae (the bright star Vega) and Cygnus x-1 lies southeast of it (in the marked position). Diagram after T. Kiang.*

Niandao is a star group on the Cygnus–Lyra border, east of Vega, and Cygnus x-1 lies in this area.

The age of Cygnus x-1 is generally estimated by astronomers as much older than the 600 years of the Chinese record. For instance, after a black hole has formed, it takes thousands of years for the companion star to evolve and increase its size so as to feed matter on to the black hole, and it takes thousands of years after that to circularize the black hole's orbit from its presumably very eccentric one. Therefore, disappointing though it is, unless astronomers are wrong in their estimates, the astonishing happening of 1408 seems to refer to another event, and not the formation of the black hole in Cygnus x-1.

SS433 and W50

Is there firmer physical evidence to connect black holes to supernovae? The closest connection between a black hole and a supernova remnant arose out of a radio astronomy study. David Clark and Jim Caswell, both then at the Commonwealth Scientific and Industrial Research Organization's Division of Radiophysics in Australia, had suspected for a long time that there were more radio stars than normal near to supernova remnants. They had used the Molongolo and Fleurs radio interferometers to map the southern radio supernova remnants and were intrigued by the number of point sources nearby to them. Clearly more than would be expected if they were quasars, these radio stars could have been radio emitting stars in the general Milky Way – or they could represent a new kind of radio emitting stellar supernova remnant.

The statistics were very hard to tie down, and Clark decided to look at individual examples. One promising radio source was inside the brightest arc of a faint supernova remnant called w50. Later, more sensitive observations of the supernova remnant showed that in fact the radio source was plumb in the middle of w50 – the two objects were clearly connected with one another. What made Clark sure of that before the evidence was clear was that he had moved to work at the Mullard Space Science Laboratory in Surrey with Fred Seward on X-ray observations with the Ariel V satellite and Seward had, in a very patient and painstaking piece of work, sorted out a very complicated region of the sky in Aquila into separate X-ray sources. One of these, A1909+04, lay over the point radio source in w50. Later X-ray observations of this source showed no pulsations and there is no evidence that it is a neutron star. Was A1909+04 the first member of a long-awaited new class of X-ray and radio sources in supernova remnants?

On a visit to Australia in 1978 to use the Anglo-Australian Telescope, Clark asked Jim Caswell for the precise radio coordinates of the radio star. He and Paul Murdin pointed the telescope at these coordinates and there lay a little rhombus of stars on the telescope finder screen, pointing to three stars at its centre, almost as if directing the astronomers to observe them. The first two were boringly ordinary, but the third had an optical spectrum indicating quite clearly that it was a binary star. Unknown to its re-discoverers, the star was in fact already catalogued under the name of ss433.

SS433

ss433 is the 433rd entry in a catalogue by N. Sanduleak and C. Stephenson of stars having emission-line spectra. Such stars have a spectrum consisting of an underlying essentially black-body component – the optically thick surface layer of a star – plus emission lines coming from an optically thin gas surrounding the star. The stars are catalogue-worthy because the gas represents some interesting interaction between the star and its neighbourhood.

The interaction causing an emission-line spectrum may have any of many causes. Some of the stars in the ss catalogue are rotating so rapidly that centrifugal force at their equators counteracts their attractive gravitational force, so that the stars shed an equatorial disc of gas which shows as the emission-line region. Some, like ss433, are binary stars with matter flowing around and between the two stars. Since hydrogen is the most common element in the Universe, the stars in the ss catalogue typically show hydrogen

emission lines (the Balmer series, the strongest line of which is H-alpha at 6563 A). Helium lines (from the second most common element) are also typical.

ss433 languished as one star among many hundreds in the ss catalogue until its rediscovery in 1978. Spectra taken then show the Balmer series and spectral lines from neutral helium, both arising from the circumstellar material ripped from one star by its companion. There were also present in its spectrum other lines which were unidentified, particularly because they seemed to come and go sporadically.

It was only after Bruce Margon of the University of Washington had monitored the spectrum of ss433 for several months that the pattern became apparent. The unidentified lines appear in pairs, the most prominent of which shift cyclically in position about a wavelength which is somewhat to the red of the H-alpha line. The cycle has a period of 164 days and the range over which each member of the pair shifts is more than 1000 A. The lines are antiphased. With this clue to the interpretation of the two strongest of the unidentified lines, the other unidentified lines could then be paired off. They show similar behaviour, oscillating in antiphase about wavelengths which lie to the red of the other Balmer lines and the neutral helium lines.

Thus the emission line spectrum of ss433 consists of three components: stationary sets of Balmer and helium emission lines arising from the interaction of a double star, and two antiphased oscillating sets of moving spectral features associated with the Balmer and helium emission, and arising from a new phenomenon. The wavelength shifts are so large that the bluer of the moving H-alpha features ranges in colour from yellow (its minimum wavelength closely matches the sodium D-lines near 5900 A) through orange,

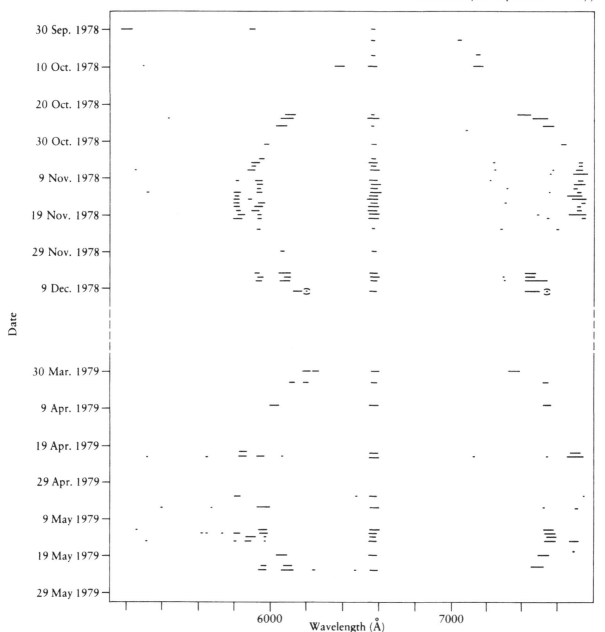

FIG. 79. *Moving lines of ss433. Observations for nine months of the wavelengths of the emission lines of ss433 were made by A. Mammano, F. Ciatti and A. Vitalone (with a break in coverage while ss433 was behind the Sun and therefore not visible at night). The stationary H-alpha line occurs at a constant wavelength of 6563 Angstroms. Either side of this are two emission lines moving in cycles with a period of 164 days.*

and the deep red of the stationary H-alpha line, to nearly 7000 A, which is beyond the red response of the eye (infrared, in fact).

The current interpretation of the moving features is in terms of a pair of equal and opposite jets of material shooting from SS433. Because of the speed of outflow of material, the spectral emissions of hydrogen and helium gases in each jet are Doppler shifted from their rest wavelengths, one (the approaching jet) generally to the blue and the other (the receding jet) generally to the red. The jets precess like a spinning top in a conical motion with a period of 164 days and so the features move in antiphased cycles of this period.

Curiously, even at the moments when the jets are in the plane of the sky, with no component of speed towards or away from us, there is a shift. It has its origin in the phenomenon of relativistic time dilation. The fast-moving hydrogen atoms in the jets have 'clocks' – the natural time scales of atomic phenomena – which are running slow with respect to ours and so emit H-alpha photons of lower frequency. The speed of the material in the jets is an astonishing one-quarter of the speed of light!

The jets of SS433 in space

Interest was focussed on SS433 because it was identified with an X-ray star within the supernova remnant W50. SS433 lies plumb at W50's centre. This lent support to D. Clark's idea that SS433 is the stellar remnant of the same supernova explosion which gave rise to W50. Perhaps SS433 contains a black hole produced in a supernova

explosion on one of a pair of stars. In this model, the black hole accretes material expelled by its companion (this material gives rise to the stationary emission lines which earned SS433 its place in the SS catalogue). As it approaches the boundary (event horizon) of the black hole, in-falling material is compressed by the intense gravitational field of the black hole. It is heated to billion degree temperatures and yields X-rays which may be the reason why SS433 is an X-ray star. The accretion seems to be at such a rate that the black hole cannot swallow all the material. The pressure of the radiation generated by heating the material may be the power-house driving the relativistic jets.

Not only is SS433 an X-ray star, it is also a radio star. Looking at the radio star with the Very Large Array of radio telescopes in New Mexico has resolved the radio image of SS433 into a blobby, elongated structure in which the blobs move out in a cone, with transverse speeds of 0.0088/arc seconds/day – matching the 0.26c jet speed at the 16 000/light year distance of the star. More than this, the directions in which the jets point – roughly east–west on the sky – align with two 'blisters' or 'ears' in the radio shell of W50 which are presumably created by the pressure of the jets on the inside of the supernova remnant. Thus the radio observations confirm the deductions of optical astronomers about the existence of the precessing jets which they invented to explain the Doppler shifts in the spectra of SS433. They firmly link the supernova remnant W50 with SS433, the supernova's stellar remnant. This constitutes the best observational evidence that supernovae produce black holes.

14

Final chapter

Some say the world will end in fire,
Some say in ice,
From what I've tasted of desire,
I hold with those who favor fire,
But if it had to perish twice,
I think I know enough of hate
To say that for destruction ice
Is also great
And would suffice.

ROBERT FROST

The escape velocity is a concept which can be applied to the constituents of the whole Universe as well as to a star or planet within it. All matter in the Universe was subject to the explosion of the Big Bang and may have been given the velocity of escape from the gravitational pull of the rest of the Universe. If this is so, then the energy of motion of the Universe, its kinetic energy, is larger than its gravitational energy and the explosion which began the Universe will never end: the Universe will continue to disperse for ever. If, on the other hand, the Big Bang was not powerful enough to overcome the mutual gravitational attraction of all parts of the Universe, the explosion will eventually coast to a halt and the Universe will collapse; when it gets small enough, it may re-explode and bounce, oscillating indefinitely.

In the technical jargon on the subject, if the gravitational energy of the Universe exceeds the kinetic energy, the Universe is closed and will collapse; if its gravitational energy is less than its kinetic energy, the Universe is open and will expand forever.

Open or closed?

There are two direct lines of attack on the problem of deciding between these possibilities.

The first consists of looking back in time at distant galaxies so far away that they represent the Universe as it was a significantly long time ago, and trying to see what the expansion rate of the Universe was then. The expansion rate may be slowing down so quickly that we can tell whether the Universe will decelerate to a stop and collapse.

To make this method work, astronomers must first measure the cosmic expansion rate of a distant group of galaxies and then determine their distance. It is here that the difficulties arise. If galaxies were all of known intrinsic brightness, like cepheid variable stars, cosmologists could use their apparent brightness as a distance measurement. In the past 20 years, several attempts using this method to determine whether the Universe will expand forever have marginally favoured the result that it is closed and will ultimately collapse. But soon after the Big Bang, the galaxies all formed at about the same time, so that distant galaxies are younger than nearby ones (because as we look farther away we look back in time). If younger galaxies are brighter than older ones (because they contain large numbers of bright stars), they will seem nearer than they really are. Hence we will be measuring the expansion rate for distant galaxies as though they were nearby, and overestimating the amount of deceleration of the Universe. These results thus seem to be overestimates of the deceleration and may be biassed in saying that the Universe is closed.

Another way in which the distance of far galaxies can be obtained is to look at their angular size. This method, applied to clusters of galaxies and to radio galaxies, has given values of the deceleration which are on the borderline between closed and open Universes and are tantalizingly equivocal. They marginally favour the open Universe.

Perhaps a better method of determining whether the Universe is open or closed is to attempt directly to estimate the kinetic and gravitational energy of the Universe to see which is bigger.

In attempting to add up all the mass in the Universe item by item to calculate its gravitational energy, astronomers have come up against the problem that they simply do not know enough about what kind of material predominates in the Universe. Most of the mass of which they are cognizant is in the form of galaxies, and the mass of an average galaxy can be measured in two ways. Astronomers can look at the speed with which stars in the outer parts of a galaxy orbit its

centre, and estimate the mass required to deflect stars by the amount observed (just as the mass of the companion star to Cygnus x-1 has been estimated by looking at its effect on the visible star HD 226868). Alternatively, they can look at the speeds with which individual galaxies deflect each other when they are situated in a cluster of galaxies.

The former method suffers from the disadvantage that if there is a halo of very faint, undetectable stars surrounding the galaxy, these have no effect on the motion of the visible stars nearer the galaxy's centre and they remain undetected. Possibly this is why the two methods for estimating the mass of an average galaxy give differing answers. Somehow, astronomers may be missing most of the matter in an average galaxy!

What form this missing matter takes has been the subject of much speculation. None of the observations exclude the possibility that between or around galaxies lie enormous masses of black holes, faint red dwarf stars, rocks or hot gas (at temperatures of around 1 million degrees). There is not much cold hydrogen, for this would be seen to play a larger part in absorbing the light from distant quasars than it in fact does. Discoveries by X-ray satellites of X-rays from the vicinity of clusters of galaxies has shown that between galaxies in the clusters there does exist very hot gas (at 100 million degrees), but probably not in such abundance that it can close the Universe.

Instead of looking at the components of the Universe item by item and adding them all up to determine the gravitational energy of the Universe, Allan Sandage has attacked the problem by looking at how faithfully the nearby galaxies follow the Hubble law, that their red-shifts are proportional to their distances from us. He argues that where galaxies do not follow this law closely, the departures from the law are caused by local clumps of matter (other galaxies, other clusters of galaxies or whatever) and the amount by which they are deflected from the Hubble law tells how much matter is deflecting them. Sandage obtains a result which he tersely summarizes:

Taken at face value these values suggest that (1) the deceleration is almost negligible. . . . (2) the Universe is open, and (3) the expansion will not reverse.

If Sandage is right, the Galaxy will become increasingly isolated from its neighbours as they recede from it. The Galaxy itself will ultimately cease to shine. Already a significant fraction of its mass is locked up in dark stars – white dwarfs and the end products of supernovae: black holes and neutron stars. An increasing proportion of its gas will have been processed through stars and increasing amounts of metals will be thrown back into the interstellar medium by supernovae. When the gas gets too metal-polluted, stars which form from it will not be able to shine.

Supernovae not only mark the death of individual stars, they hasten the aging of our Galaxy, possibly towards a dark, cold and lonely death as it finds itself alone in the Universe.

At a conference in Cracow in 1973, John Wheeler conducted an opinion poll of the assembled cosmologists on the question of whether they thought the Universe was closed or open. Of course, truth is not decided by democratic vote – the result of the poll only gives an indication of what most informed people think. Most cosmologists put themselves into the 'don't know' category, and were prepared to wait for more solid evidence before pronouncing on the subject. We apparently shall not for a while be able to read the final chapter in the life of our Galaxy, although it is at this moment being written sentence by sentence among the stars, and punctuated by supernovae.

Booklist

K. M. V. Apparao. The Crab Nebula. *Astrophysics and Space Science*, 25, p. 3 (1973).

P. J. Bancazio & A. G. W. Cameron (eds). *Supernovae and their Remnants*. New York: Gordon & Breach (1973).

D. H. Clark. *Superstars*. London: J. H. Dent (1981).

D. H. Clark. *The Quest for SS433*. N.Y.: Viking (1985).

D. H. Clark & F. Stephenson. *The Historical Supernovae*. New York & London: Pergamon Press (1977).

J. Danziger & P. Gorenstein (eds). *Supernova Remnants and their X-ray Emission*. Dordrecht: D. Reidel (1983).

R. D. Davies & F. G. Smith (eds). *The Crab Nebula*. Dordrecht: D. Reidel (1971).

Flagstaff symposium on the Crab Nebula, *Publications of the Astronomical Society of the Pacific*, 82, p. 375 (1970).

R. N. Manchester & J. H. Taylor. *Pulsars*. San Francisco: W. H. Freeman (1977).

S. Mitton. *The Crab Nebula*. London: Faber & Faber (1978).

M. J. Rees & R. J. Stoneham (eds). *Supernovae: a Survey of Current Research*. Dordrecht: D. Reidel (1982).

D. N. Schramm (ed.). *Supernovae*. Dordrecht: D. Reidel (1977).

I. S. Shklovsky, *Cosmic Radio Waves*. Cambridge: Harvard (1960).

I. S. Shklovsky, *Supernovae*. London: Wiley (1968).

J. C. Wheeler (ed.). *SN I*. Austin: University of Texas (1980).

Index

Lightning Source UK Ltd.
Milton Keynes UK
UKOW07f2040040516

273569UK00005B/237/P

9 780521 189798